The Rubens Garden

Klara Alen

THE RUBENS GARDEN

A Masterpiece in Bloom

HANNIBAL
RUBENSHUIS

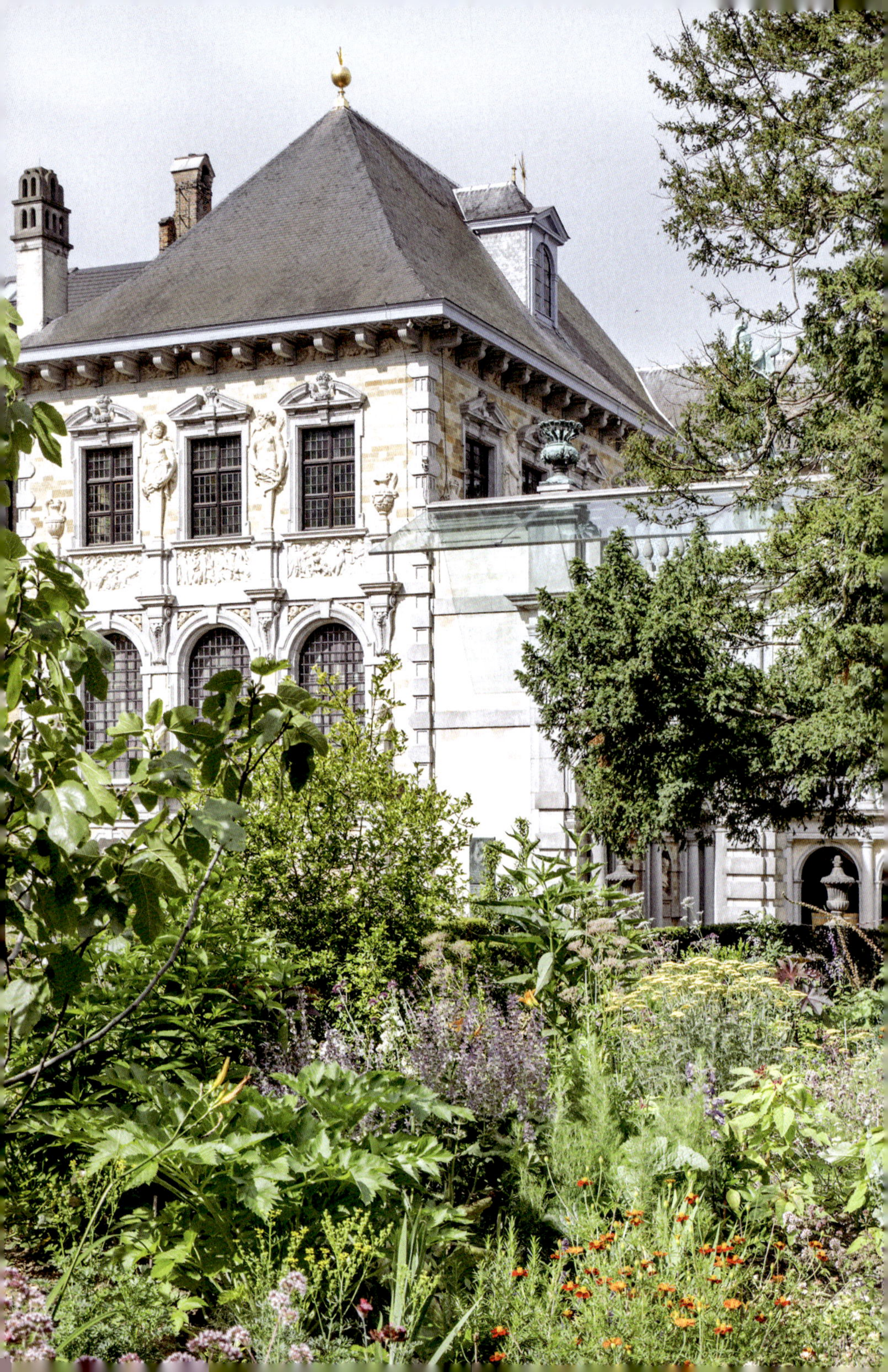

With heartfelt thanks to Dr med Bettina Leysen,
without whose generous support this publication would not have been possible.
The English translation of this book was made possible through
the kind support of the American Women's Club of Antwerp.
The Rubenshuis is indebted to Boortmalt, its principal sponsor.

BOORTMALT
> MASTERS OF MALT

**Also, remind
Willem the gardener
that he is to send us
some Rosile pears and figs
when there are some,
or any other delicacy
from the garden.**

Rubens's only written reference to his garden,
dated 17 August 1638

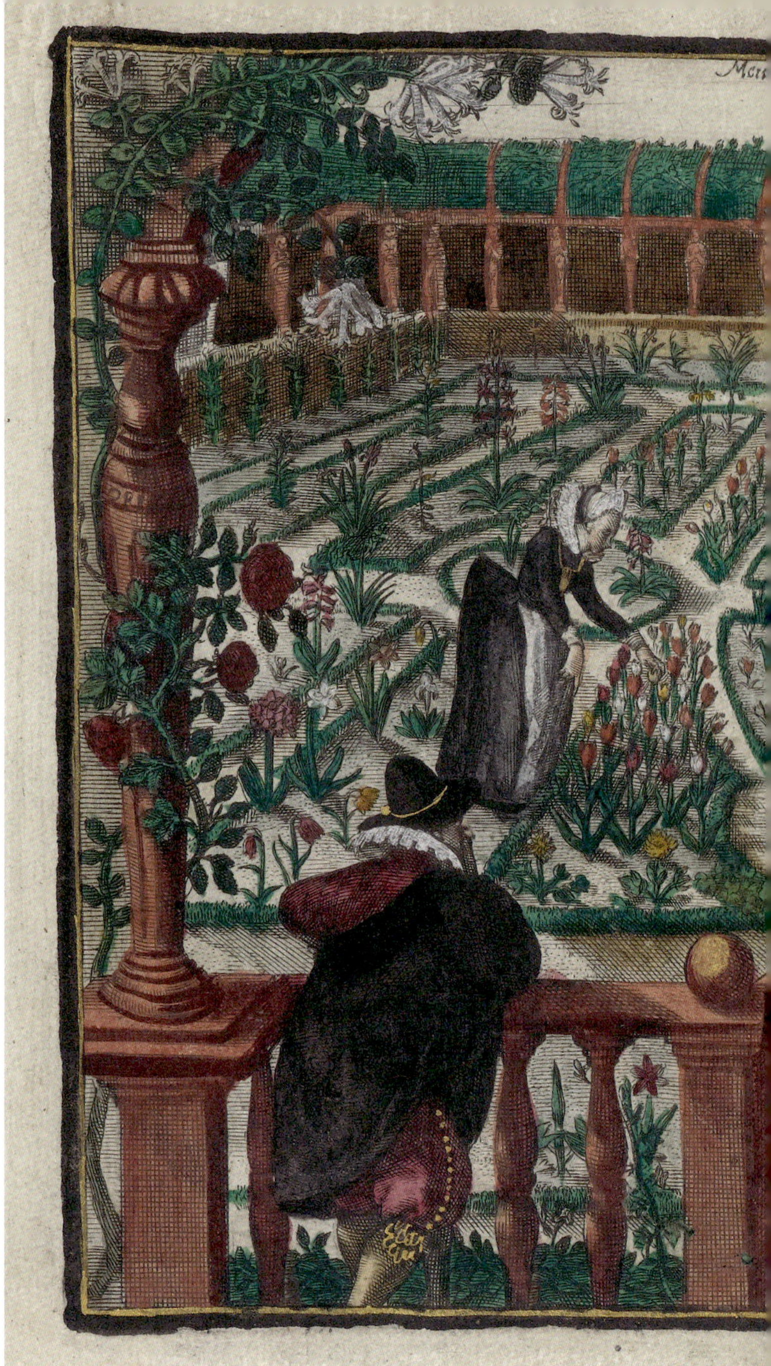

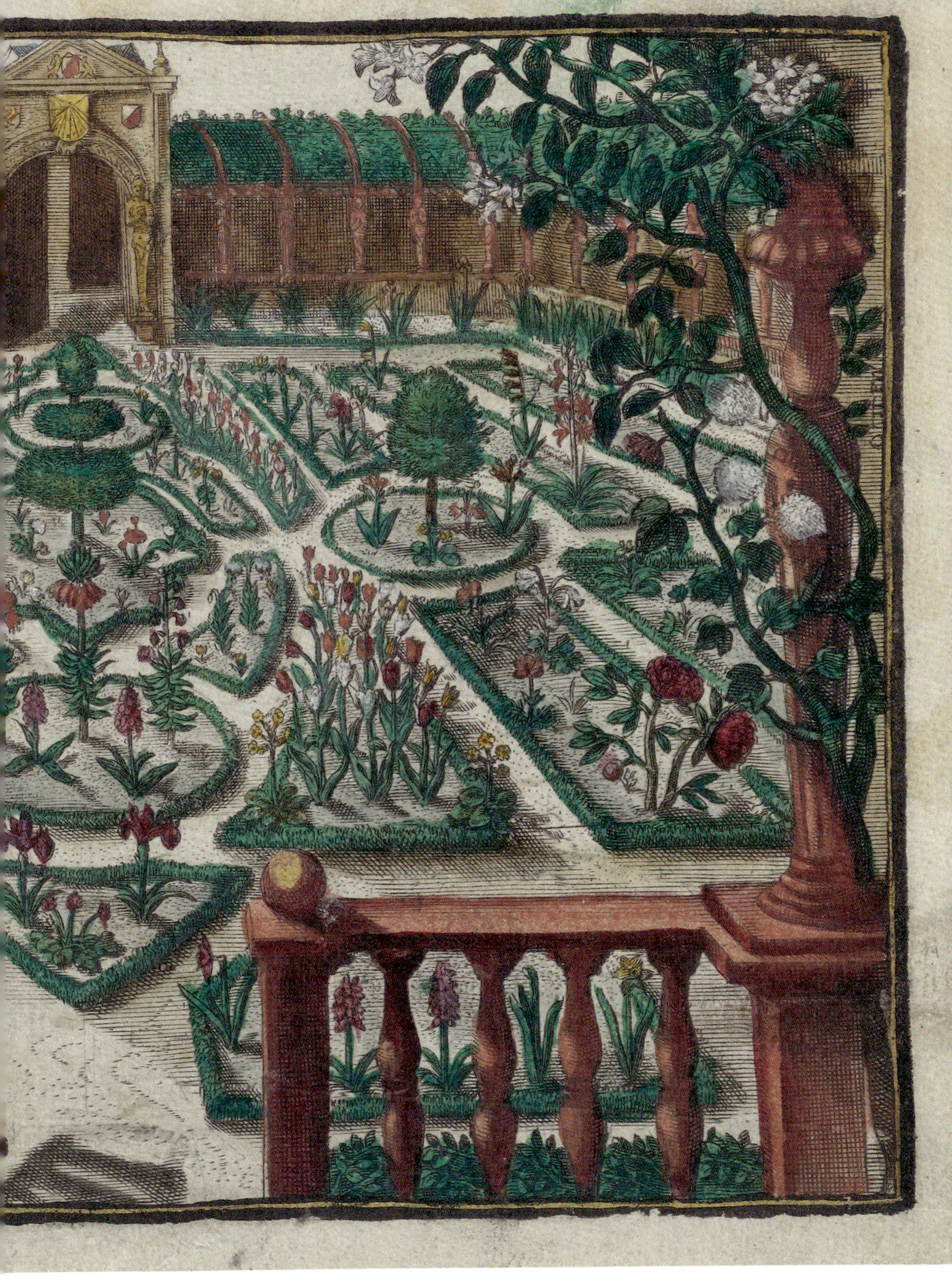

Crispijn de Passe II, *Spring* from the *Hortus Floridus*, 1614, Oak Spring Garden Foundation, Upperville, Virginia (USA)

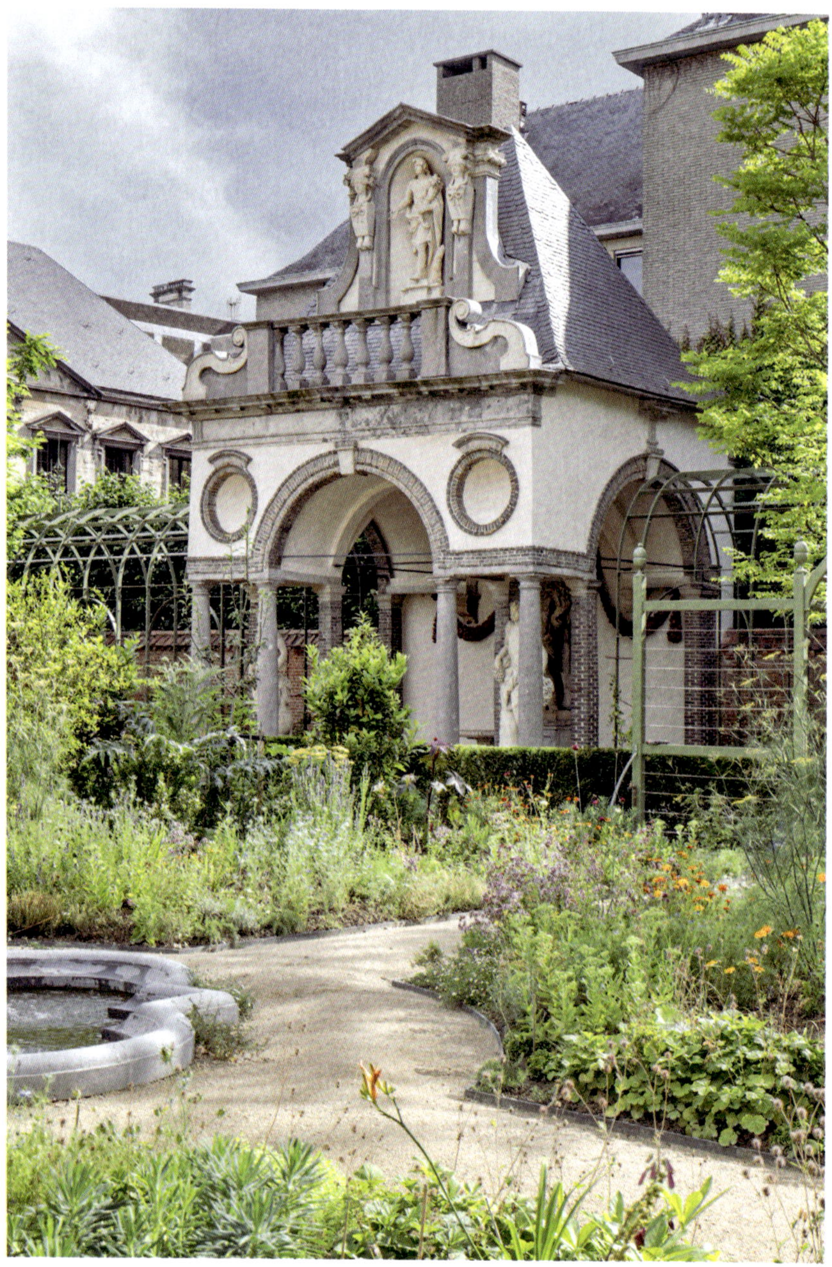

The garden pavilion in the summer of 2024

Foreword

Planting a garden, by definition, is an act of optimism, of belief in growth and development, of faith in the future. At the same time, the new garden of the Rubenshuis is deeply rooted in the site's centuries-old history. Thorough scientific research underpinned the design, which meticulously restores, preserves, and enhances the *genius loci*, its spirit of place. In a sense, this new-old garden has become a 'seventeenth-century island' in the heart of Antwerp.

Its use of historical dimensions based on the Antwerp foot (a unit of length that was smaller than the imperial foot), the restoration of central and radial sight lines and perspective depth, the clever proportioning of paths and parterres, the addition of shrubs and trees of varying dimensions, and the berceau with arcades that is proportionate to the existing historical architecture: the new-old garden evokes Rubens's unique and highly Italianate contribution to the garden tradition of the Low Countries on every level. Archaeological excavations also added a turbulent chapter to the history of Antwerp.

Gardens as witnesses of the cyclical return of seasons question our linear conception of time, of past, present, and future. The cosmic order and the cycle of seasons is immune to our annual record-keeping, after all. A good garden thus becomes a virtually timeless place, a plant-based clock without hands. A place where time is slower, where years are counted in the invisible growth rings of trees, and where flowers set the agenda. It creates an island in time, in the heart of the city.

A good garden, even an historic one, is always here and now. A place that sharpens our awareness, putting past injuries, present-day concerns, and future ambitions into perspective. This garden already did all of the above for Rubens: it was a canvas to refresh his and our troubled minds. The keystone above the central, depressed arch in Rubens's garden portico depicts the head of Medusa, said to ward off evil. Those who enter here leave behind evil intentions, burning ambition, and negative emotions.

This garden keeps the world out more than ever. Ars Horti's designers radically reversed the principle of the *borrowed landscape,* turning the gaze inward to reinforce the illusion of an island in time. However, attention to the past in this eternal now also involves thinking about the future. While the garden may visually appear to be a *hortus conclusus,* an enclosed garden, the design also takes major societal challenges into account. Below ground, we harvest geothermal energy and large amounts of rainwater, which slowly infiltrates into the ground. Besides being an ecological stepping stone and a place where you can take refuge from the heat, this garden also contributes to the climate resilience of Antwerp's city centre.

Those lucky enough to work in a place like this soon realise that their stewardship is very temporary, that their role is nothing more than a small link in a centuries-long chain of care. Rubens experienced professional triumphs and private grief and happiness in this very spot for 30 years. Both his wives, Isabella Brant and Helena Fourment, loved to relax in this garden, marvelling at how their children had grown while watching them play. Rubens portrayed them among honeysuckle and roses. The artist is buried just a few hundred metres from his beloved garden, in the chancel of St James's Church. Today, church bells still measure how the light and seasons sweep across his good earth. With the garden, our largest museum gallery, and this publication, we hope to make our guests more keenly aware of this place and its history, and of the short, heavenly seasons that are ours to enjoy. *Il faut cultiver notre jardin.*

Bert Watteeuw, Director of the Rubenshuis

CONTENTS

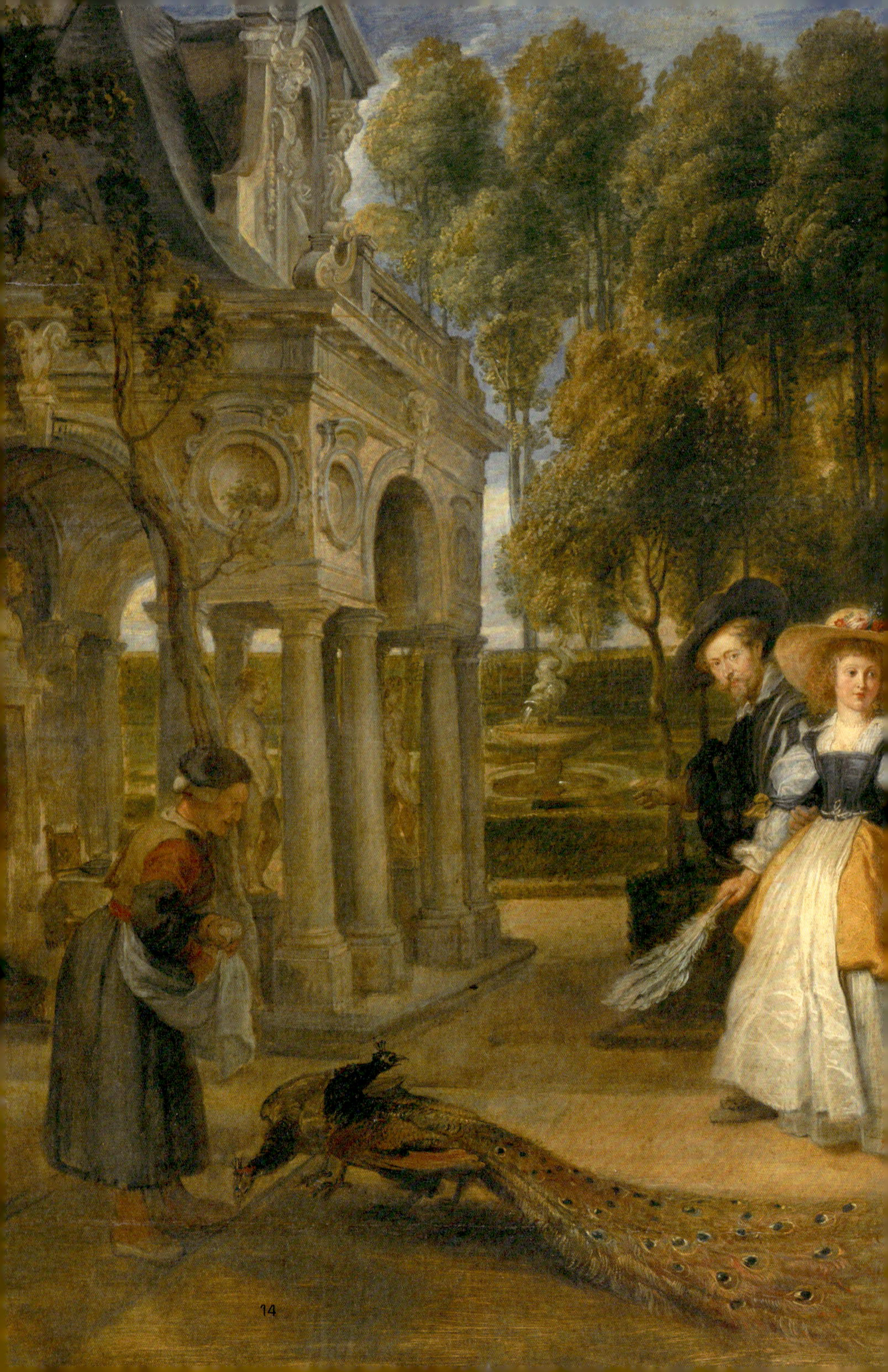

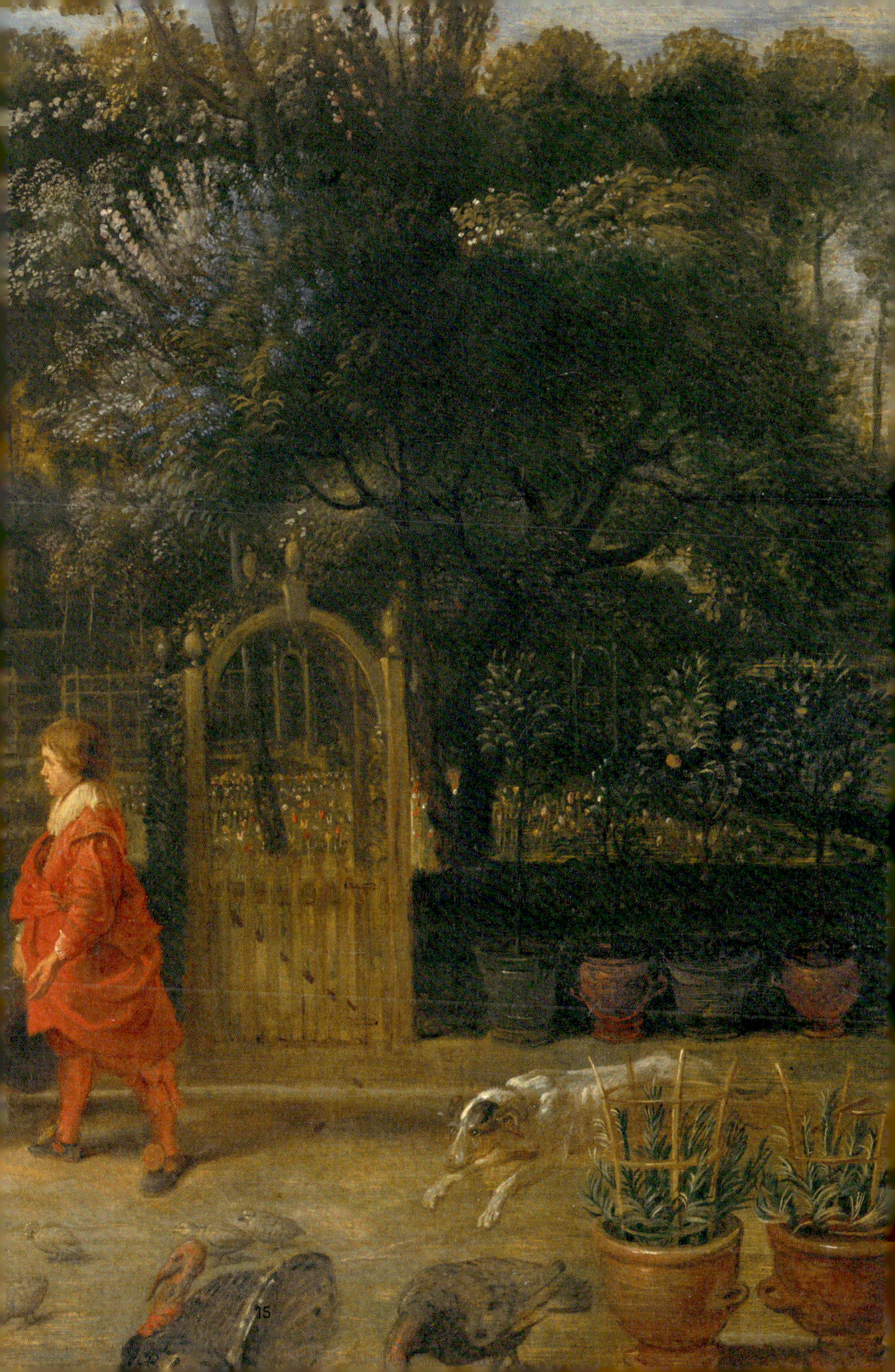

PREVIOUS SPREAD Peter Paul Rubens (atelier), *Peter Paul Rubens, Helena Fourment and Their Son Nicolaas Walking in Their Garden ('The Walk in the Garden')*, c. 1630–1631, Bayerische Staatsgemäldesammlungen – Alte Pinakothek, Munich

HISTORICAL RESEARCH FOR THE RUBENS GARDEN

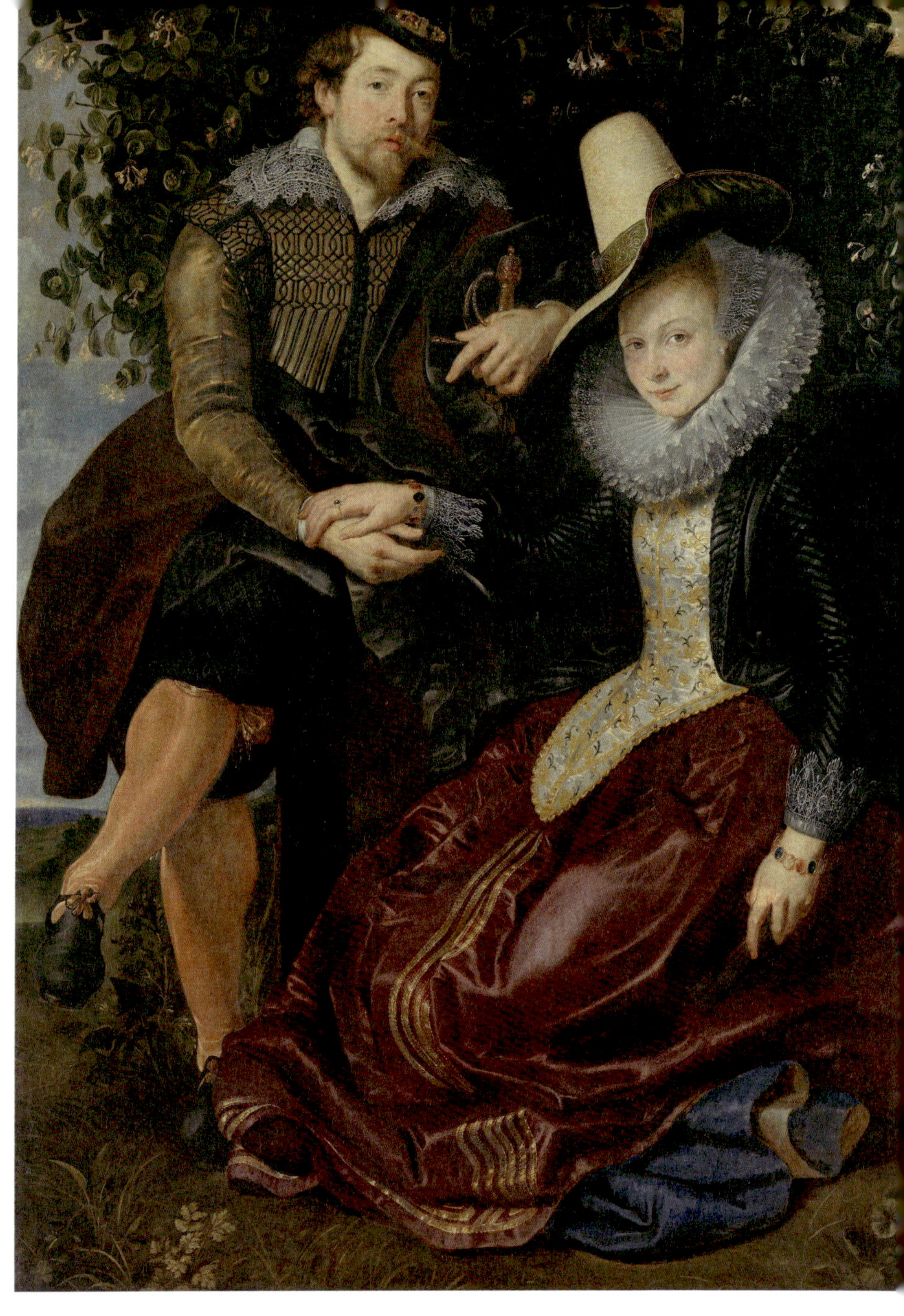

Peter Paul Rubens, *The Artist and His First Wife Isabella Brant in the Honeysuckle Bower*, c. 1609, Bayerische Staatsgemäldesammlungen – Alte Pinakothek, Munich

Embellishing the garden

On 1 November 1610, a notary finalised the sale of a house with a former bleachfield along the Wapper in Antwerp. The new owners of this exceptionally large and green plot on the fringe of the city centre were Rubens and his wife Isabella Brant, and they had a very special reason for wanting to live here. Upon her return from Germany in 1587 – after the death of her husband Jan Rubens – Rubens's mother Maria Pijpelinckx had moved in with her parents, who lived in *Den Cleynen Arnold* on the Meir (where it still stands at no. 54 today), with her three young children. From the roof window of his grandparents' house, the young Rubens could see the rows of wooden frames on the bleachfield, on which cloth was stretched to dry. Rubens would have also seen a row of tall linden trees to the left, adjacent to the bleachfield and next to a long garden wall. This was the vast garden of the Kolveniersgilde, the Antwerp guild of arquebusiers whose members met here for shooting practice and festive gatherings.

Rubens knew the surroundings like the back of his hand because he used to play here as an 11-year-old boy. More than 20 years later, he saw the plot's great potential: it was big enough to build a large studio, convert a spacious existing house, and enjoy a large, sun-drenched garden in the company of his family and guests. The family relocated here in the autumn of 1615, with their two small children, Clara Serena and Albert. Work on the house and studio was still in progress, and there was no formal garden when they moved in. According to an Antwerp merchant, Rubens had already spent 24,000 guilders on his house by 1618, an eye-watering amount in those days. He was referring to the extravagant redecoration of Rubens's home and the construction of the studio, as well as the addition of a portico as a folly. As is often the case with renovations, the garden came last and was probably completed around 1620. Like today's gardens, Rubens's garden would change constantly in the following decades.

There were lots of sources of inspiration to choose from for the garden design and plantings. Isabella Brant's childhood home on Kloosterstraat had a large garden. That is also where Rubens painted his self-portrait with Isabella Brant in a bower with a honeysuckle bush. On their wedding day, 3 October 1609, they were still able to take in the last *Lonicera* of the season in the garden before the blooms faded. During his eight years in Italy, Rubens had also visited the most remarkable city gardens in Genoa and the fashionable gardens of palazzos and villas in Mantua, Florence, and Rome. But closer to home, there were other exceptional gardens on which he could feast his eyes. Thanks to his appointment as court painter, Rubens had access to the vast gardens of the Archdukes Albert and Isabella at Coudenberg Palace in Brussels. The most extraordinary and fashionable species of flowers of the time bloomed there, but their gardens were also home to a toucan, lions, tigers, monkeys, porcupines, peacocks, camels, and ostriches, among others. Specimens from this Brussels zoo are featured several times in the work of Rubens and his contemporaries, including his colleague Jan Brueghel I. Rubens's circle of friends and clients in Antwerp also included a number of garden lovers and collectors of unusual plants, such as Mayor Nicolaes Rockox.

ABOVE RIGHT Jan Brueghel I, *The Archdukes Albert and Isabella in the Garden of Coudenberg Palace in Brussels*, 1620s, Rubenshuis, Antwerp

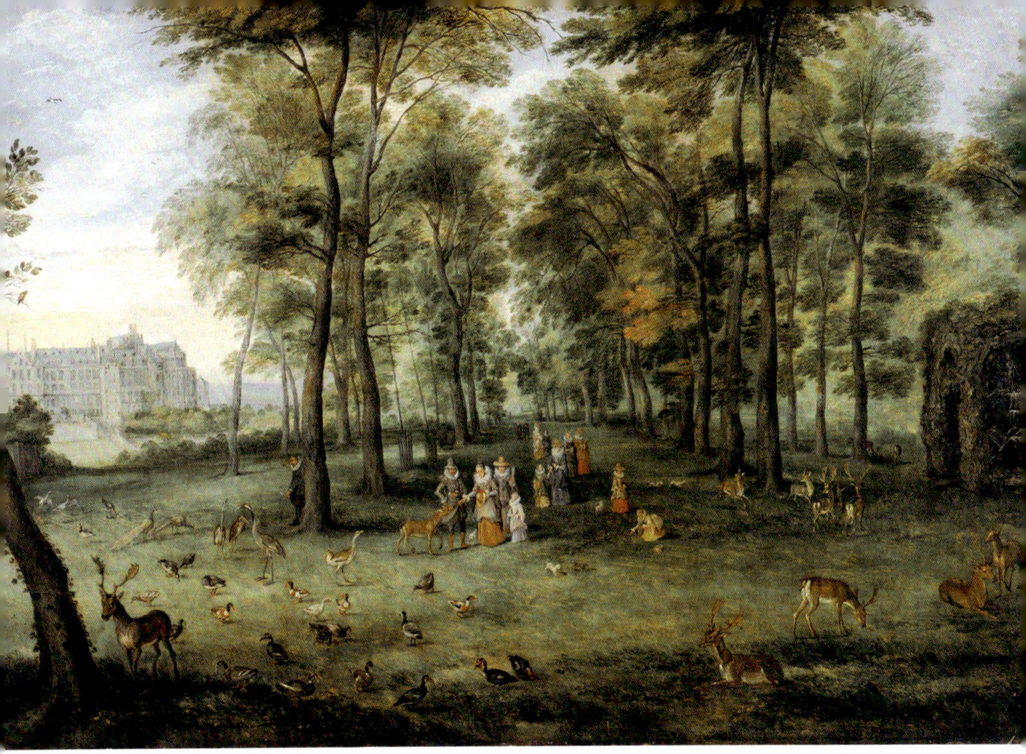

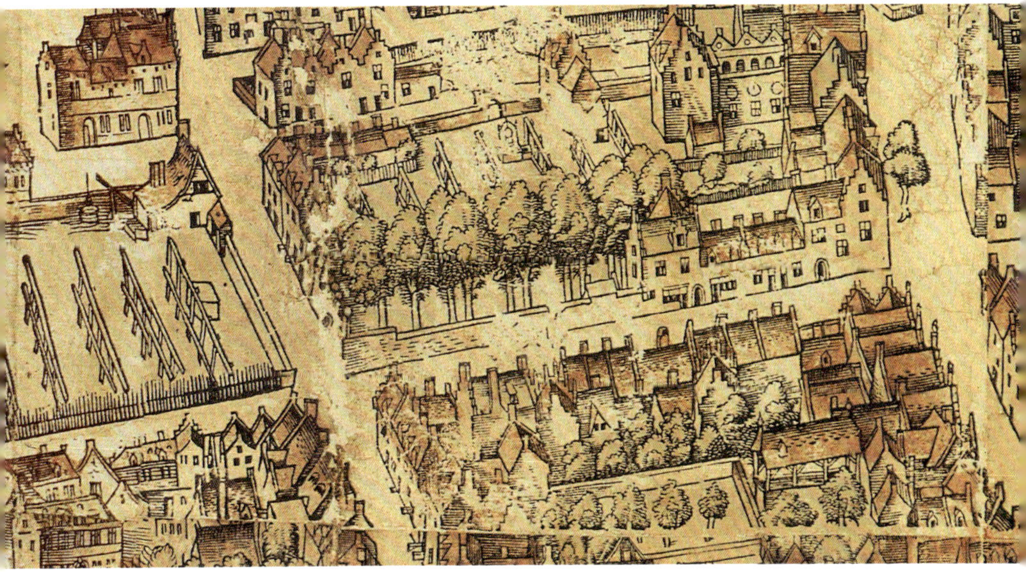

BELOW Detail of the later Rubenshuis site with the wooden frames on the bleachfield and the Kolveniershof with the double row of linden trees (centre), from: Virgilius Bononiensis (draughtsman) and Gillis Coppens van Diest (printer), *Map of Antwerp*, 1565, Museum Plantin-Moretus, Antwerp

Rubens's garden became a meeting place for his family and friends. A playground for his children Clara Serena, Albert, Nicolaas, Clara-Johanna, Fransje, Isabella, Peter Paul junior and Constance. The workplace of Willem and Jaspar, his gardeners. A prestigious networking venue for the artist's international clients. A garden of love for Rubens and Isabella Brant and later – following his second marriage – Helena Fourment. A backdrop for Rubens's own work and that of contemporaries such as Jacques Jordaens, David Teniers, Anthony van Dyck, and Gonzales Coques. The territory of the Rubens family's cat. A picking garden for their scullery maid Willemyne. An exhibition space for a peerless collection of antique sculptures. A showroom with a harmonious Italian portico and a charming garden pavilion. A drying area for large canvases and panels. A safe haven for three horses. A heavenly place to linger, enjoy, read, feel reenergised, make music, unwind, and dance, where the senses were stimulated and the mind invigorated. 'As the painter sharpens his tired eyes from prolonged and concentrated gazing with a mirror or among greenery, thus our dulled and distracted minds are refreshed in the garden', the scholar and humanist Justus Lipsius wrote in his book *De Constantia* (On Constancy), in which he revisited the ideals of Greco-Roman stoicism.

Il acheta donc une grande maison dans la ville d'Anvers, il la rebastit à la Romaine, & en embellit les dedans, qu'il rendit commodes pour un grand Peintre & pour un grand Amateur des belles choses. Cette maison estoit accompagnée d'un jardin spatieux, où il fit planter pour sa curiosité des arbres de toutes les especes qu'il put recouvrer.

He thus bought a large house in Antwerp, refurbished it in the Roman style and embellished the interior, making it comfortable for a great painter and lover of beautiful things. This house came with a large garden, where he had trees planted of all the species he could find to assuage his own curiosity.

Roger de Piles, Rubens's biographer, *Conversations sur la connaissance de la peinture*, 1677, based on the *Vita Petri Pauli Rubenii* by Philip Rubens, the artist's nephew.

Willem and Jaspar, Rubens's gardeners

Digging, sowing, growing, planting, fertilising, sweeping, raking, hoeing, weeding, grafting, propagating, sanding and painting trellises and gates, showing guests around the garden, harvesting. These are just a few of the many tasks listed in 17th-century garden maintenance contracts in the Low Countries. We do not know whether Rubens also concluded such a comprehensive contract with his gardeners. Around 1640 – the year of his death – the family employed three women and five men. Coachman Jan used the stables in the garden. Willemyne the scullery maid knew best what to pick each season in Rubens's garden. But Jan Drion (the manservant, who also laid out the table), Franchoys (who ground the master's pigments) and Anneken and Adriaenken (the maids) also knew their way around the garden. Gardeners Willem Donckers and Jaspar Verbrugghen tended to the garden.

Willem took care of the general upkeep of Rubens's garden and – along with Jan Drion – was the highest-paid employee. His annual salary was 72 guilders, a high sum that can be explained by the relationship of trust between Willem and Rubens. In August 1638, Willem had access to the garden while Rubens and his family stayed at their country house *Het Steen* in Elewijt near

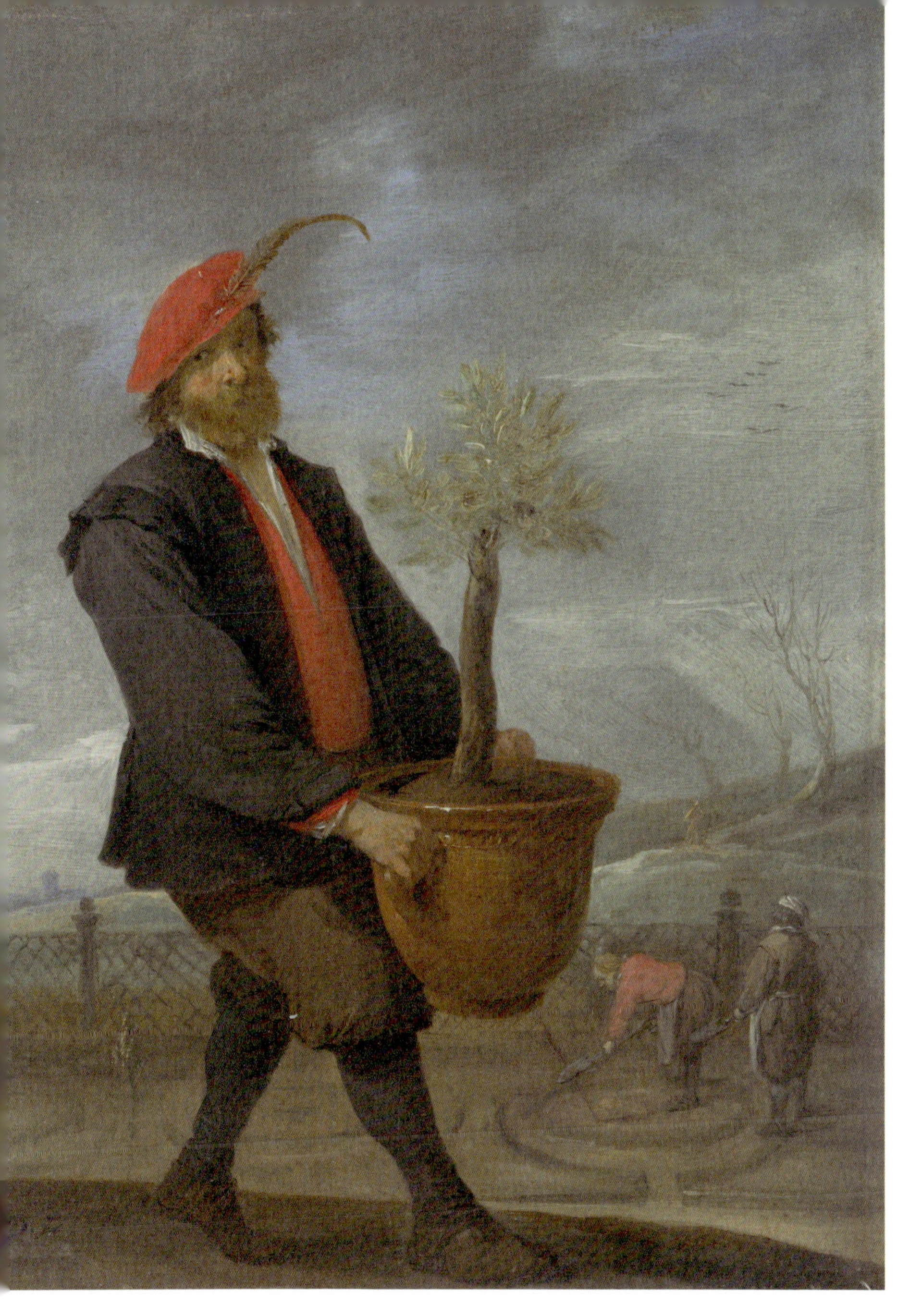

David Teniers II, *Spring*, c. 1644, The National Gallery, London

Mechelen. No doubt gardeners also continued to work in the garden when Rubens and his family moved to an inn in Laeken for a year and a half from August 1624 because the plague was rampant in Antwerp. Without maintenance, the Antwerp garden would have been completely overgrown. This trusting relationship between Rubens and Willem was significant for another reason: theft of flowers – in particular of precious (tulip) bulbs – or of the garden's spoils, such as eggs, was rife. It was thus in Rubens's interest that Willem kept an eye on it.

Jaspar was paid by Rubens to look after the orange trees. He must have been a sought-after gardener who worked with assistants or subcontractors because Jaspar also did gardening jobs on the plot of Rubens's neighbours, the Kolveniers, and worked for the Moretus family, the successors of printer-publisher Christopher Plantin. He supplied fruit trees to the Moretuses and took care of garden maintenance, the pruning of the grapevines, and the pickling of the grapes. Moreover, Jaspar became a close neighbour of Rubens during the last years of his life. On 30 September 1637, he bought *De Blauw Banck,* a house with a large garden on

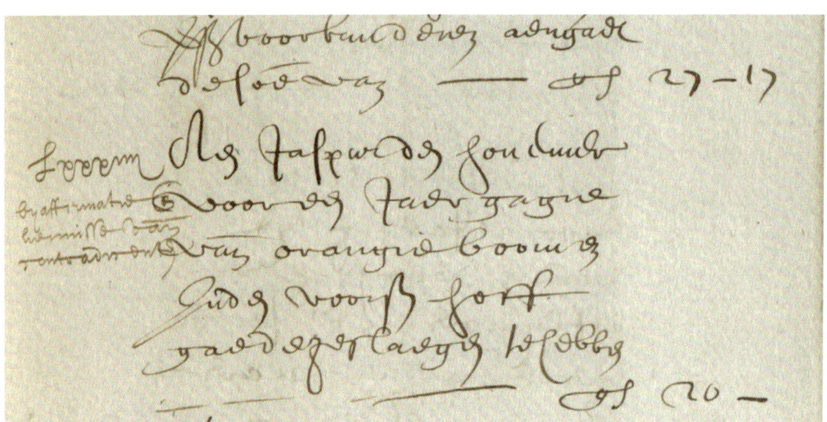

Payment to Jaspar the gardener for tending to the orange trees in the *Staetmasse* or settlement of Rubens's estate in 1645, Felixarchief, Antwerp

the nearby Hopland. Flowers would go on to play a crucial role in the lives of the Verbrugghens for at least two more generations. Jaspar's son, Jaspar Pedro, was a flower painter and owned orange trees, laurel trees, and tulips. The grandson of Rubens's gardener incorporated flowers in his paintings and designs for tapestries.

The gardener's bothy can be found in the south-west corner of the current Rubens Garden. This small edifice was built during the first restoration of the Rubenshuis (from 1937 until 1946) by city architect Emiel Van Averbeke and leading Antwerp garden architect Georges Wachtelaer. The cottage was used to store the orangery plants. Van Averbeke and Wachtelaer found inspiration for the building and the overall design of the garden in a print of the garden made by Jacob Harrewijn in 1692, more than half a century after Rubens's death. Although Rubens's gardeners would have let citrus plants overwinter in the cellar, attic or in a heated indoor space, Willem and Jaspar would have also had a place in the garden where they could store their gardening equipment.

Estate inventories of Antwerp contemporaries of Rubens give an idea of which garden tools they would have used. The garden tools of Antwerp alderman Charles de Tassis, for instance, included distillery equipment, but also three iron shovels, two digging spades, two metal rakes, an iron lever, a parrot with its aviary, and two small glass flowerpots. The Duarte family, Rubens's neighbours whose house stood on the Meir had, among other things, a copper watering can to water the garden, several painted sticks for the garden, two wooden stands for trees, and also a cockerel, two hens, and a chicken coop. Antonette Wiael, the widow of art dealer and panel-maker Jan van Haecht, kept 21 tubs with carnations and a copper watering can that was used in the garden in her front attic on the street side. In his home

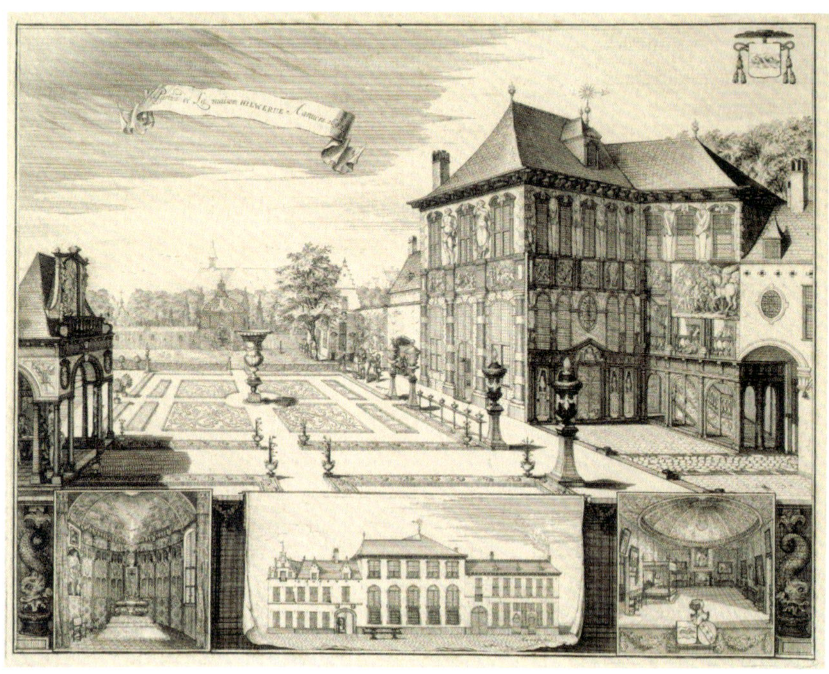

opposite St George's Church, Master Jacques Cornelissen owned a brass watering can that was used in the garden. The popular gardening manual *Den Nederlandsen hovenier* by Jan van der Groen includes a print depicting the most important gardening tools. The print with *'hof-gereetschap'* shows a tree lopper (A), a caterpillar nest picker (B), a grafting knife (C), a rake (D), a pruning chisel or debarking tool (E), a garden hoe (F), a knive (G), a saw for grafting (H), a hatchet (I), hand scissors (K), an iron chisel (L), a wooden mallet (M), a hand trowel (N), a spade (O), a garden fork (P), and a shovel (Q).

LEFT Jacob Harrewijn after Jacob van Croes, *The Rubens House at Antwerp, ('Parties de La Maison Hilwerve à Anvers')*, 1692, Rubenshuis, Antwerp — RIGHT Jan van der Groen, *Van't Hof-gereetschap* from *Den Nederlandsen hovenier*, 1696, Museum Plantin-Moretus, Antwerp

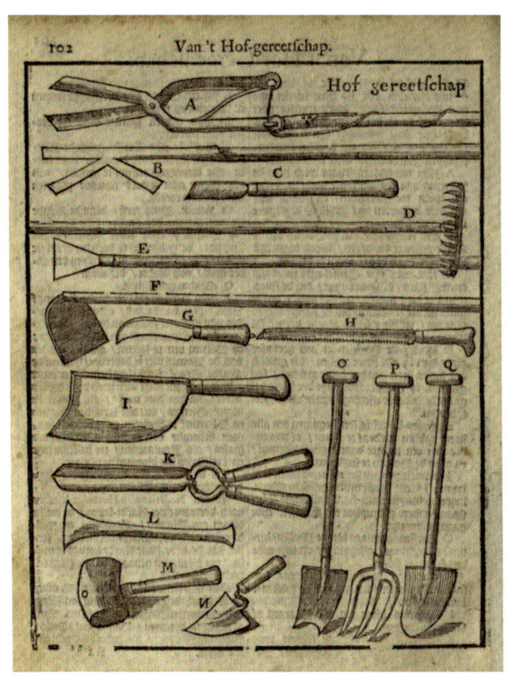

The wise gardener does not let any time of the year go to waste. At the start of the new year, he takes out his gardening tools to start gardening and search for and gather herbs. He sharpens his pruning shears, folding knives, hatchets, chisels, shovels, and spades, and he makes sure he has all the tools he needs for farming and gardening. Because as they say, good tools are half the job.

FVS, *Den verstandighen hovenier*, 1672

Atro telluris occulta. Hidden in the darkness of the earth

'Voracious time threatens all that is hidden in the darkness of the earth', 13-year-old Albert, Rubens's eldest son, wrote in a book about Roman coins. Chances are he had no idea what lay beneath his family's garden. During the recent redevelopment of the Rubens Garden, soil was removed up to 3.5 metres below ground level with a view to installing water catch and infiltration basins. For Antwerp archaeologists, this was a unique opportunity to delve deeper into the darkness of the earth.

In the northern and southern areas of the garden, the city's archaeologists found hundreds of hornpits and goat phalanges in different locations, proof that tanners worked on this site in the late Middle Ages. Goatskin leather was a luxury product in those days. This very fine, soft leather was used among other things for gloves. Animal hides were delivered from the countryside to tanners in the city with the horns and legs still attached, as a kind of quality label. The hides were placed in oak tubs filled with a pickling solution made of ground oak bark mixed with water. They were then dried, scraped, sanded, and rolled. The tanner

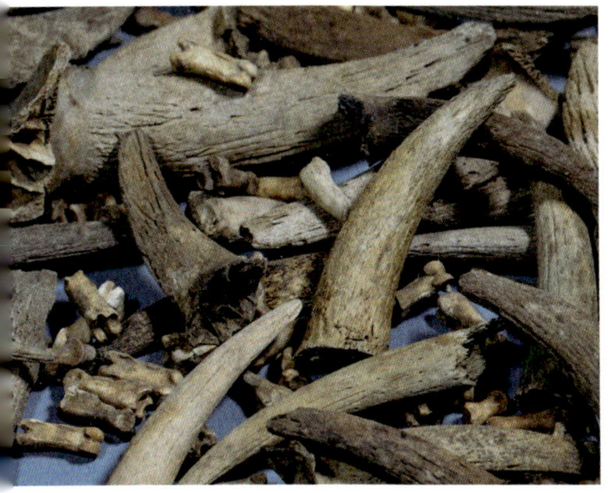

sold the outer layer of the horn – the horn tissue – to the horn worker. Animal horns were used to make many things, including musical instruments. The horn's core (the airy and brittle part of the bone or hornpit) and the phalanges were thrown into a waste pit. Archaeologists found several of these large waste pits in the Rubens Garden.

In the southern part of the garden, the archaeological survey unearthed five square brick pillars in 16th-century masonry. The evenly-spaced pillars formed a semicircle. It soon became clear that these were the remains of a large Calvinist temple with a diameter of 16 metres. Remnants of the temple had already been found during previous archaeological excavations, led by Georges Wachtelaer, in the Rubens Garden almost one hundred years ago. The structure dates from 1566, the year the Iconoclasm erupted in Antwerp during which 200 Calvinists stormed the cathedral, destroying its interior in a matter of hours. During the night, members of the Kolveniersgilde succeeded in securing their own chapel in the cathedral. The next morning, they moved their altarpiece to the town hall. They also took the statue of their patron saint, Saint Christopher, to the Vrijdagmarkt, and later to their shooting range next to the present-day Rubens Garden. The statue was so high that the guild needed a pulley to hoist it over the garden wall.

In an attempt to smooth ruffled feathers after the Iconoclasm, Protestants were granted permission to erect six temples within

Hornpits and goat legs after cleaning by the Department of Archaeology of the City of Antwerp

the city walls. One of these temples was built on the plot of the later Rubens Garden, away from public streets and enclosed by walls, as had been agreed to by the Antwerp city council. Construction started on 24 September 1566 in the southern part of the garden. The temple had an octagonal (*achtcantich*) floor plan with bluestone (to a man's height) and wooden walls. The roof was covered with slate roof tiles. The art historian and illustrator Joris Snaet created a reconstruction for the Rubenshuis based on historical research.

A contemporary chronicle describes how 'how people worked very diligently on this church and how some Calvinist men and women sometimes even worked without pay'. The funds came from 'upstanding citizens and noblewomen of Antwerp', as well as citizens of Tournai, Armentières, and Valenciennes. To speed up construction, they had contributed large sums of money and also donated jewellery. A contemporary account describes how 'a Catholic who was drunk' somehow found his way into the temple on 31 December 1566. This (Catholic) drunkard knocked over the pulpit. It broke, and the man was caught, 'if, however, a follower of Calvin or Luther would have done this, he would have paid for it with his life', the chronicler notes.

On 13 June 1567, less than a year after the foundation stone was laid, Margaret of Parma ordered that all Protestant temples be demolished and all traces of the Reformation erased. A bribery attempt by the *Vleeschouwers van Borgerhout* (butchers of Borgerhout) to use it as a butcher's hall to peddle their wares amounted to nothing. The temple was razed to the ground, 'to ensure no remnants would remain in the city of this false synagogue'.

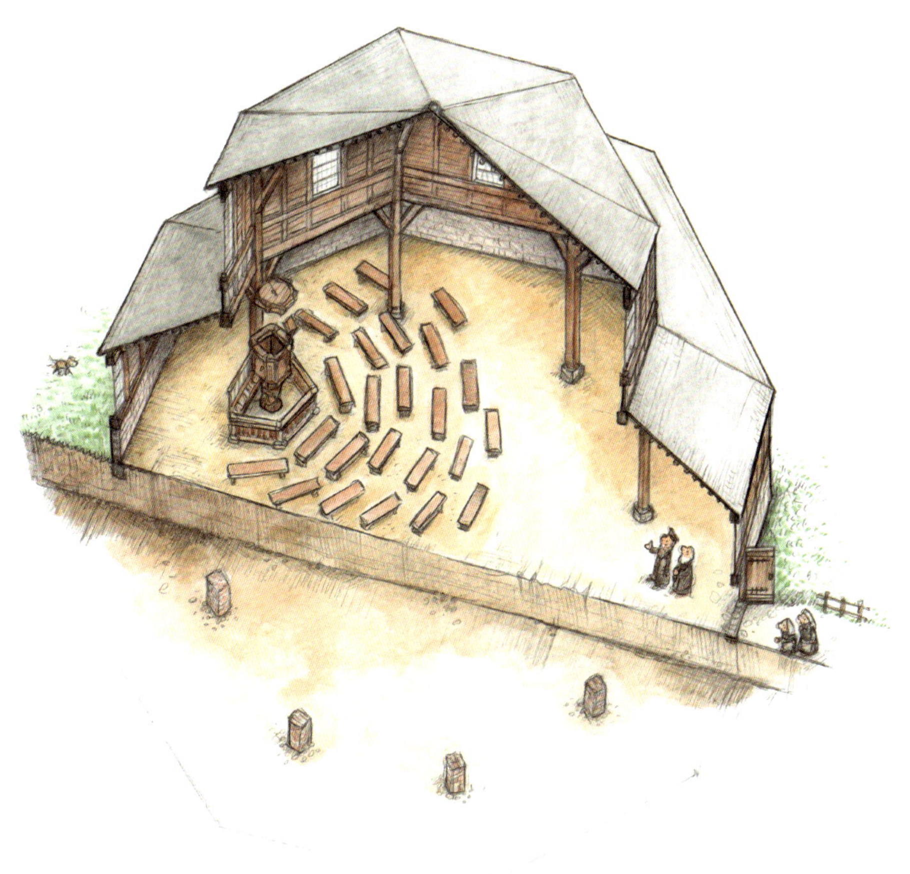

Joris Snaet, *Reconstruction, drawing of the Calvinist temple in the later Rubens Garden*

Many Flemish Protestants fled abroad. Jan Rubens, the father of Peter Paul, who was a lawyer and Antwerp alderman, was also forced to flee Antwerp in 1568. As a Protestant, he was no longer safe in the Catholic Netherlands. Together with his wife Maria Pijpelinckx and their four children, he relocated to Cologne. There Jan began an affair with Princess Anna of Saxony, the second wife of William of Orange. He was almost beheaded because of this extramarital affair, but the extraordinary generosity of his wife Maria, who stood by her husband despite everything, may have well saved his life. After two years in prison, Jan Rubens was released under house arrest in 1573. Peter Paul Rubens was born in Siegen on 28 June 1577, the sixth of seven children.

The portico: the crown jewel of the Rubens Garden

Rubens's home cum studio presumably retained its original appearance until the mid-eighteenth century, after which it underwent extensive remodelling. The only elements of Rubens's original design that were preserved are the Italianate portico and the garden pavilion. We do not know when and by whom the portico and pavilion were built. The portico may have been completed by 1621, when the structure was deemed a fitting backdrop for a portrait of Isabella Brant by Anthony van Dyck. In Rubens's design, the portico, which resembles a triumphal arch, connected the 16th-century residence with the newly-built *schildershuys* or studio, and served as a gateway to the garden. This type of garden architecture was completely new in Antwerp. In the present-day Rubens Garden, the scenography evolves like a gradual transition, as envisaged by Rubens. From the courtyard, you walk through the portico to the pavilion at the back of the garden. The orientation of the parterres, towards the pavilion, accentuates the sense of perspective and depth.

Rubens used text and imagery on the portico to develop an ingenious programme about art, nature, and the artist's virtues. Eagles bearing garlands of fruit and flowers, symbols of fertile

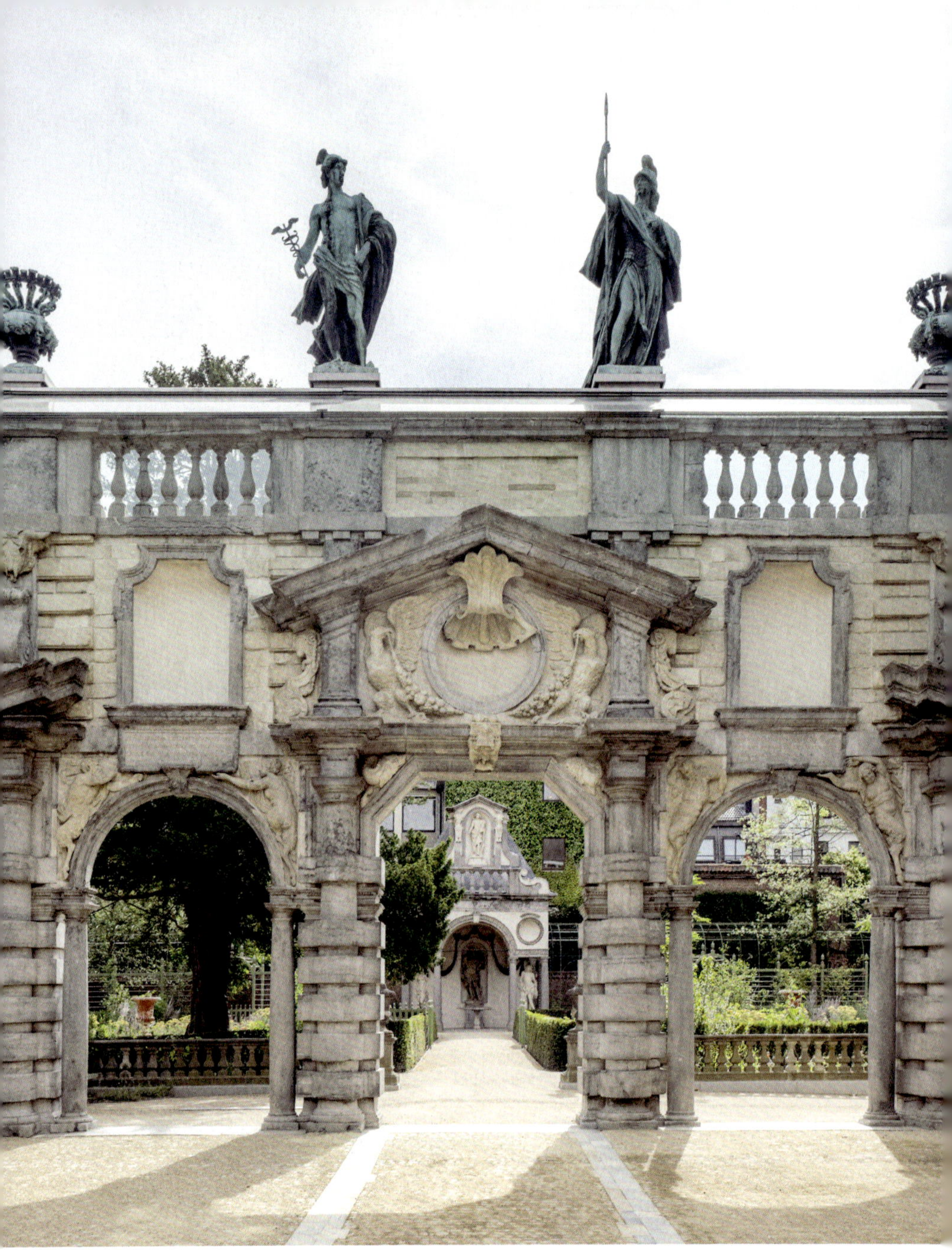

Sight line from the courtyard through the portico to the pavilion

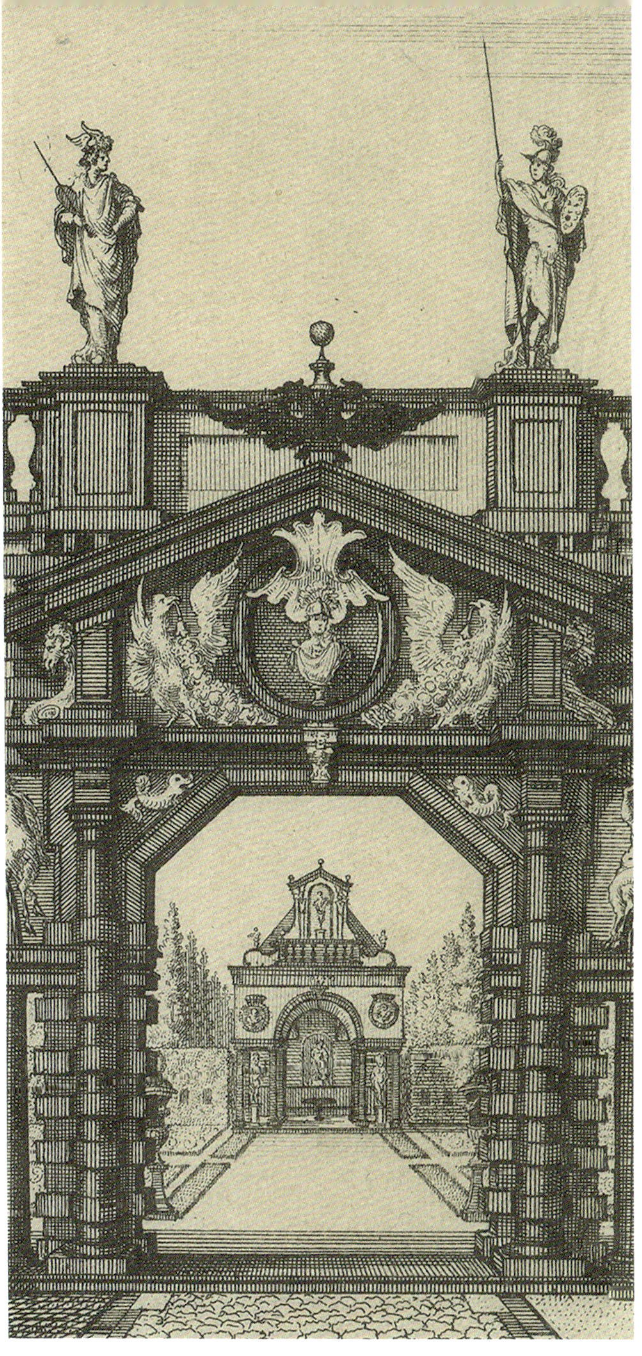

DETAIL FROM Jacob Harrewijn after Jacob van Croes, *The Rubens House at Antwerp ('Maison Hilwerve à Anvers, dit l'hostel Rubens')*, 1684, Museum Plantin-Moretus, Antwerp

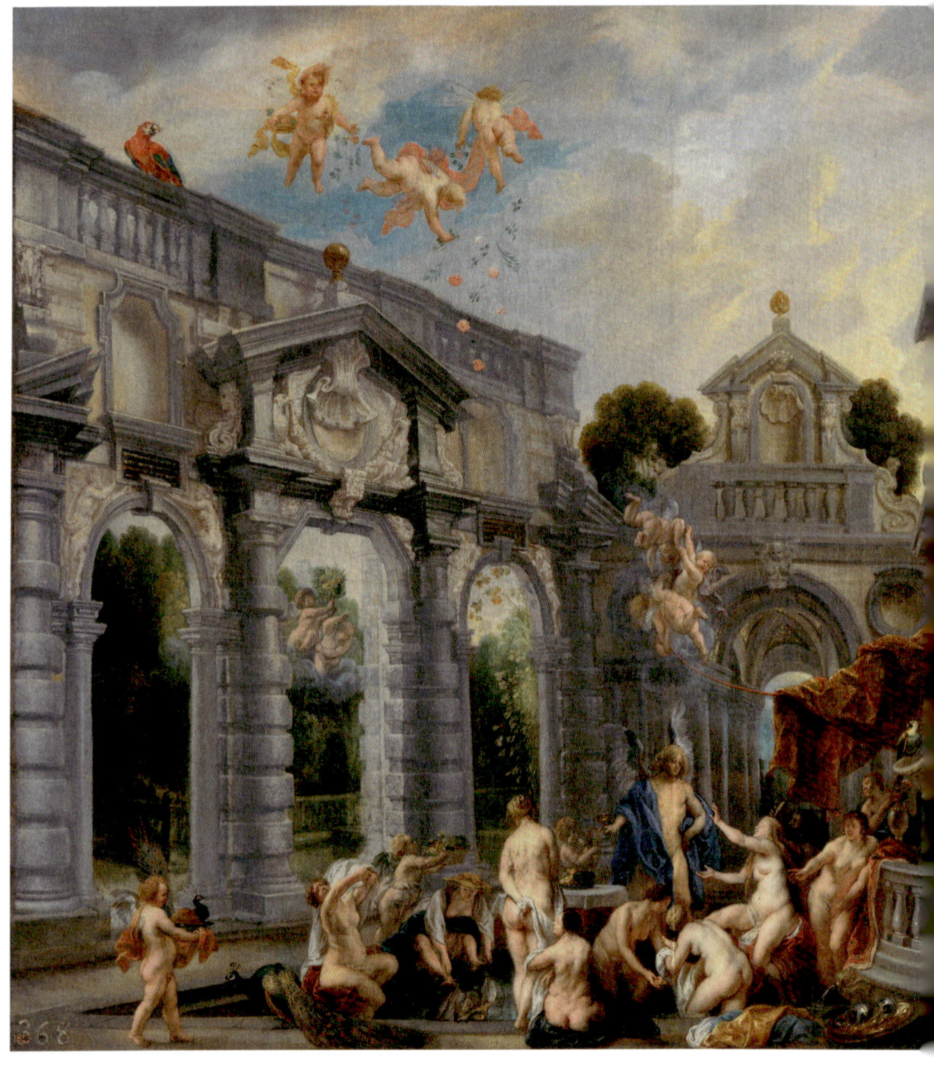

Jacques Jordaens, *Cupid and Psyche*, c. 1640–1650, Museo Nacional del Prado, Madrid

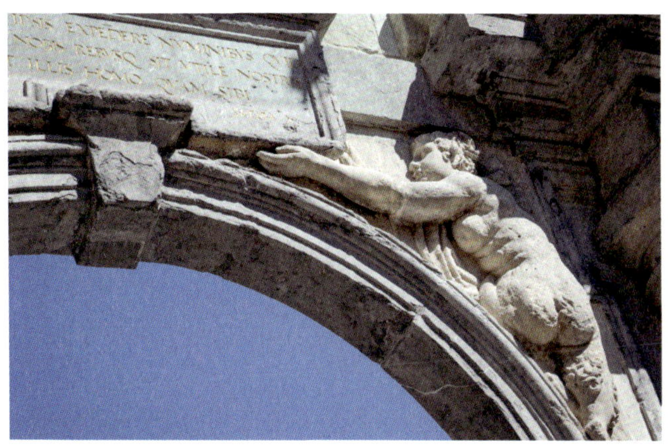

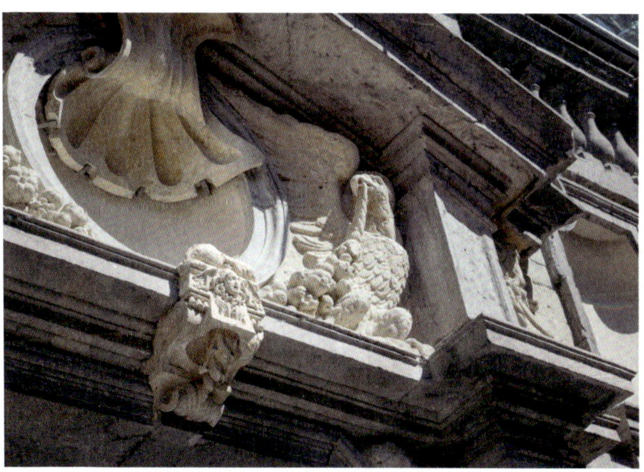

ABOVE Female satyr near the left passageway of the portico —
BELOW Keystone with Medusa's head

nature, as well as peace, harmony, and *virtuoso* or virtue can be seen in the centre. A bust of Minerva, the ancient goddess of wisdom, stood in the shell niche until the 20th century. The sculpted keystone in the shape of a winged head surrounded by snakes and lightning bolts represents Medusa, a mythological creature who was used to fend off enemies.

Satyrs lean on the spandrels of the portico's side passageways: these mythical, mischievous forest creatures are half-human and half-animal. They also have a strong connection with nature, and belonged to the retinue of Dionysus, the god of wine, fertility, and *joie de vivre*. In this instance, they bear panels with gilt lettering, inscribed with verses from the Tenth Satire by the Roman poet Juvenal.

The panels are reminiscent of the *lex hortorum* at the entrance to Renaissance gardens, a 'garden law' or code of conduct for visitors that usually included a life philosophy. This is the only

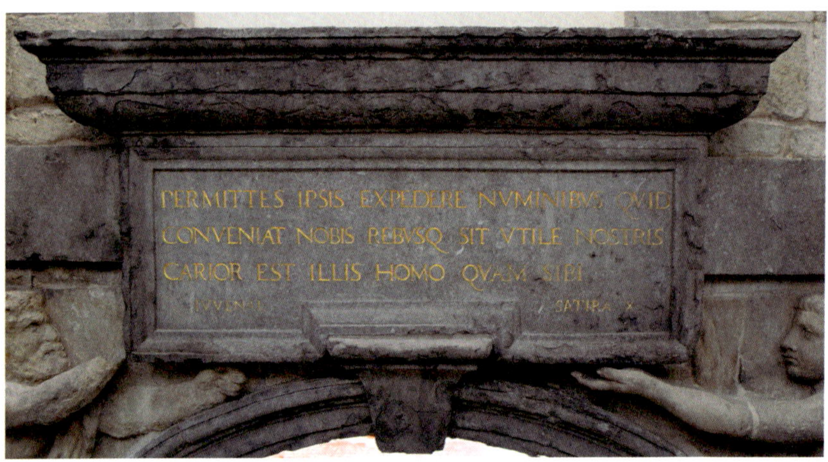

place where Rubens addresses his guests directly, in the words of Juvenal. The Latin text on the left refers to divine providence as a source of well-being: *'Permittes ipsis expedere numinibus, quid conveniat nobis, rebusque sit utile nostris carior est illis homo quam sibi'* (Leave it to the gods themselves to provide what is good for us, and what will be serviceable for our state. Man is dearer to them than he is to himself). The quote on the right urges rational self-control and fortitude: *'Orandum est ut sit mens sana in corpore sano fortem posce animum et mortis terrore carentem nesciat, irasci, cupiat nihil'* (Prayers should be made for a healthy spirit in a healthy body, for a courageous soul, which does not fear death, which is not evil and does not covet anything). With these words, Rubens exhorted visitors to his garden to behave wisely and exercise self-control. Man is happiest when he feels free in his mind and accepts fate and the will of the gods. The kind of peace and freedom you can find in... the garden.

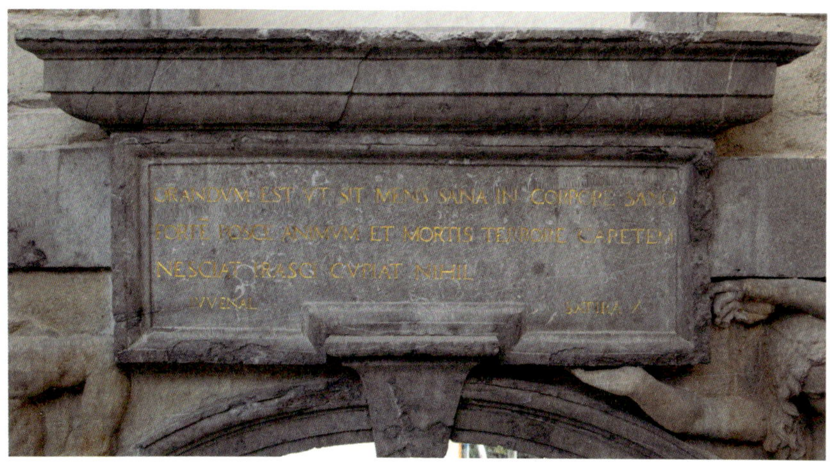

LEFT AND RIGHT Panels with verses from the Tenth Satire of the Roman poet Juvenal at the left and right passageways of the portico

The pavilion: the eye-catcher

Looking from the portico towards the garden, all eyes are on the eye-catcher of the garden pavilion: the Roman demigod Hercules, a symbol of triumphant virtue and a familiar sight in Italian Renaissance gardens. He holds the apples he got from the garden of the Hesperides, the eleventh of his Twelve Works. In the garden, Hercules can unwind. He leans on a real wooden club.

A marble table dating from Rubens's time stands in front of Hercules. Originally, Hercules was flanked by Bacchus and Ceres, the god and goddess of wine and fertility, respectively. The three statues turned this place into an environment where virtue and lust went hand in hand. In the 20th century, Ceres was replaced by Venus, the goddess of love and fertility, for reasons yet unknown. The statue of Venus was created by Antwerp sculptor Willy Kreitz (1903–1982). Honos (Honour) is perched at the top of the niche in the pavilion, enthroned on a cornucopia, a horn of plenty. He personifies the spirit of the place (the *genius loci*) and watches over the happiness of the home and garden.

In his garden design, Rubens envisioned a spectacular view of the garden through the central arch of the portico. The façade of the Italianate garden pavilion, in the form of a *serliana* (an arch flanked by two lower rectangular openings) with *oculi* (round

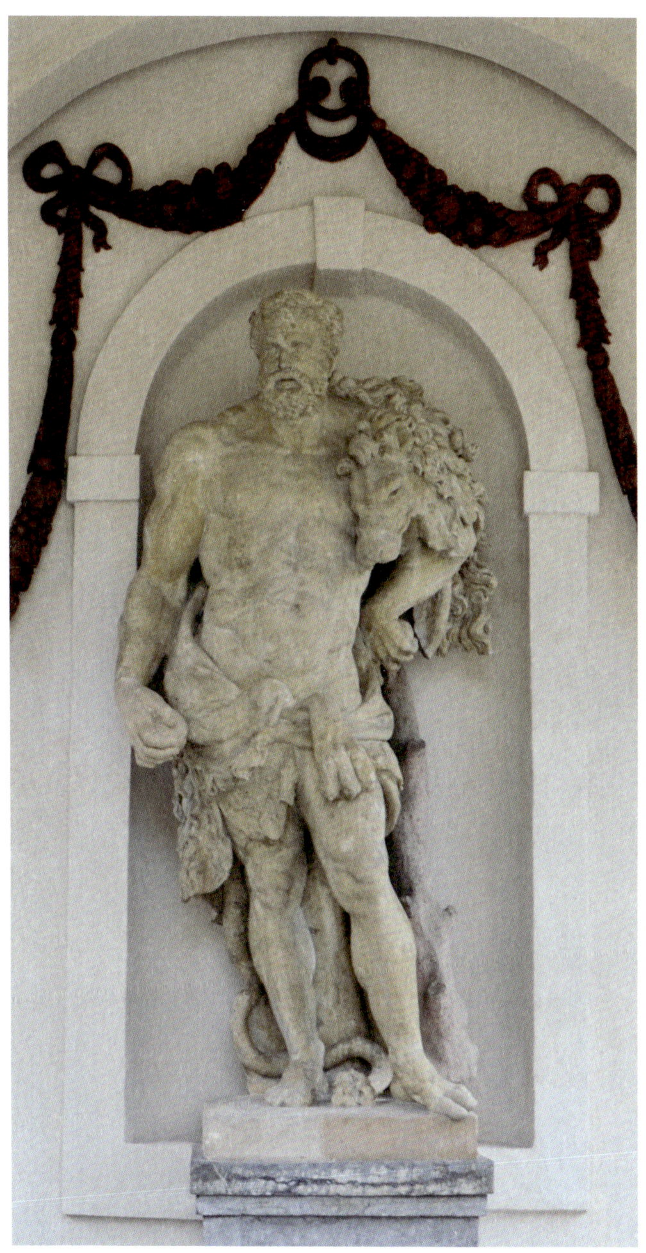

Hercules in the pavilion of the Rubens Garden

niches), needed to be visible in its entirety from the courtyard. In the new garden design, the proportions of the large arches of the garden pavilion and portico have been respected, including in the width of the central portico-pavilion axis but also in the arch width of the circular path around the garden and the berceau.

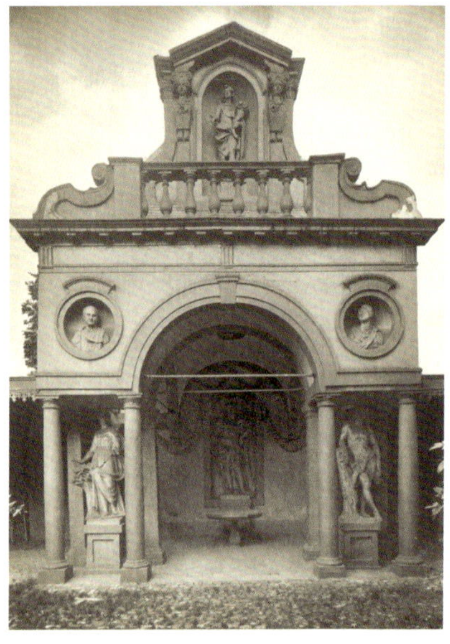

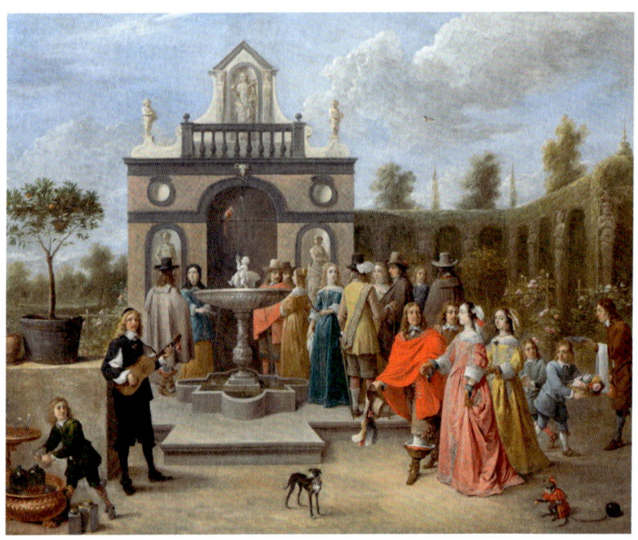

ABOVE The garden pavilion before the restoration of the Rubenshuis, early 20th century — BELOW David Teniers II, *An Elegant Company Before a Pavilion in an Ornamental Garden*, 1651, The Phoebus Foundation, Antwerp

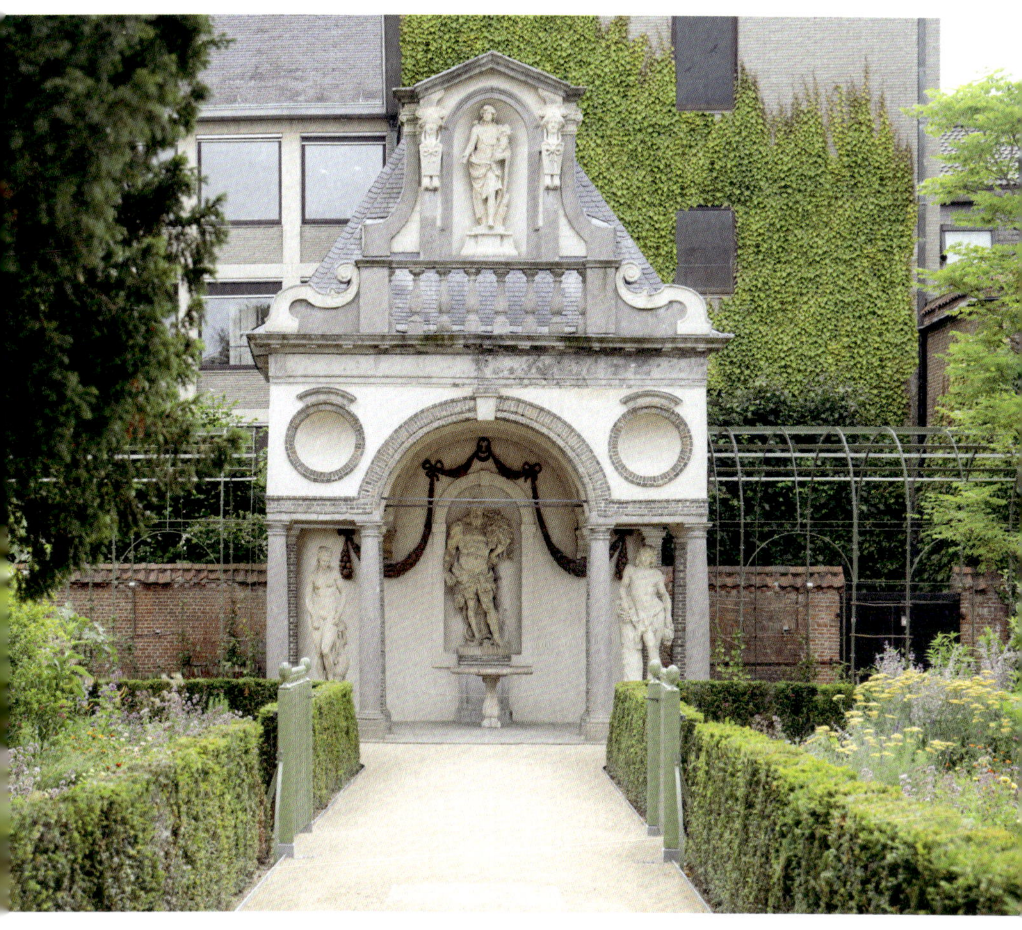

The garden pavilion in 2024

The berceau

In the new Rubens Garden, the garden pavilion designed by Rubens will be surrounded by a copper-green berceau, a covered garden path. In Renaissance and Baroque gardens, deciduous trees or hedges were encouraged to spread over a wood or wrought iron trellis and pruned to form a tunnel. A berceau looks a bit like an upside-down cradle, which explains its name (the French word for cradle). Some berceaus had openings or niches on the sides.

A work by David Teniers II depicting an elegant company in a garden features a pavilion that was clearly inspired by Rubens. On the right side of the building, a tall hedge grows with openings at regular intervals. Did Teniers get the inspiration for his painting directly from Rubens's garden? If that is indeed the case, there must have already been a berceau here just like there is today. Such a structure may also have been mentioned in documents when the property changed hands. The deed of sale to Rubens shows that the property had a *gaelderije* in addition to a large gate and entrance and a courtyard. Perhaps the word *gaelderije* was a synonym for gallery or arcade and thus the berceau. The word *loove* (foliage) does appear in a deed of sale of the plot in 1538.

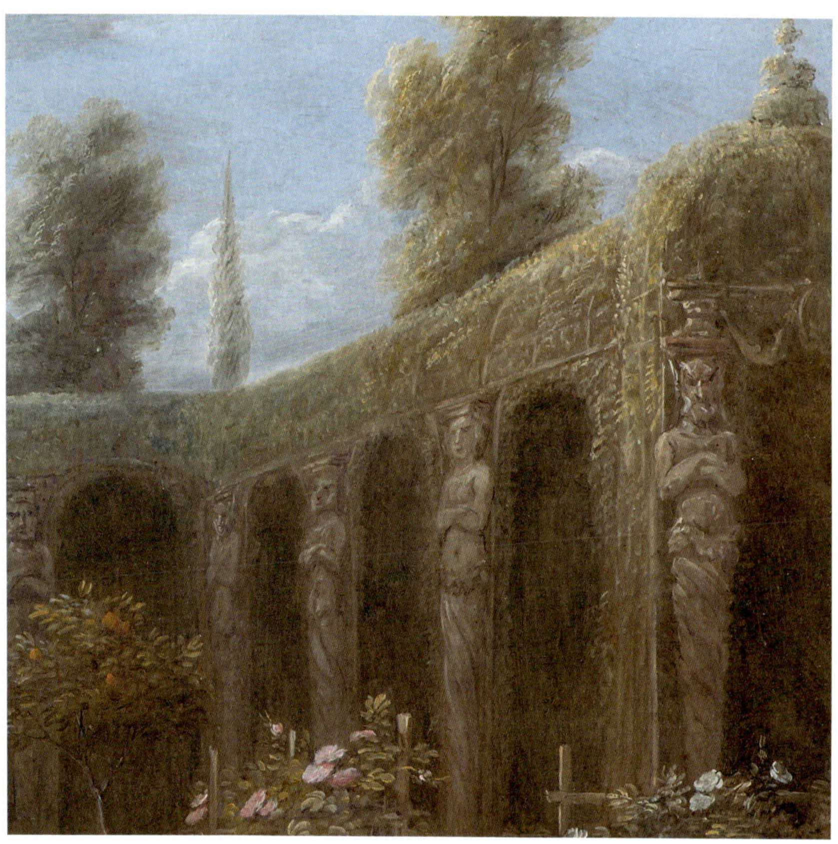

Today six oak caryatids and herms with flower vases and baskets on elegant bluestone pedestals flank the berceau. These sculptures of female and male figures were originally produced during the first restoration of the Rubenshuis from 1937 until 1946, some to a design by sculptor Willy Kreitz. At the time, they adorned the pillars of a wooden pergola that could not be restored. Unfortunately, the statues were also badly damaged and had not been publicly displayed for quite some time. They were restored

DETAIL FROM David Teniers II, *An Elegant Company Before a Pavilion in an Ornamental Garden*, 1651, The Phoebus Foundation, Antwerp especially for the new garden of the Rubenshuis. Shell-shaped rain canopies or awnings now protect the sculptures from wind and rain.

The repurposing of the caryatids and herms for the new berceau is thus a lasting tribute to the first evocation of the Rubens Garden. Look closely at Teniers's painting – and at other iconographic sources from that era – and you will also see herms standing between the niches against the berceau's framework. Although they appear to be made of bluestone, they are almost certainly made of faux stone, or painted wood.

Inside these arcades and bowers, you use slats and rungs, ledges or trellises, as you see fit, which you then decorate and paint in various colours such as red, white, blue or another colour. They can also be planted and laced with white bryony, honeysuckle, privet, rambling roses, sweet-briar, cleavers, hops, pumpkins, melons, jasmine, and other similar trees.

FVS, *Den verstandighen hovenier*, 1672

The berceau in the Rubens Garden is planted with ivy, grapevines, clematis, hops, honeysuckle, and roses, among others: their branches can be easily trained and tied back. This structure allows visitors to experience the garden even more as a green oasis in the busy city, as a *hortus conclusus*, an enclosed garden. The climbing plants with their growth control system against the south wall, the evergreen plantings, and the planned redevelopment of the Kolvenier site further enhance this feeling.

The epitaph of the humanist Philip Rubens, the artist's beloved brother, can be found among the greenery of the berceau, against the garden wall of the Kolveniershof. After Philip's unexpected death in 1611, Rubens commissioned a memorial stone for him to be installed in St Michael's Abbey, right next to their mother's, who was also buried in this church. He asked his old schoolmate Balthasar Moretus for help with the Latin inscription. The Latin text quotes Cicero, among others: 'He left us, but he has not been forgotten. He lives on through his talents and writings'. In 1833, the epitaph was lost when the church was demolished. In 1951, the stone resurfaced, however, during works in Otto Veniusstraat. It had been broken into three parts and was used as a sewer cover. In the new Rubens Garden, the epitaph has been restored.

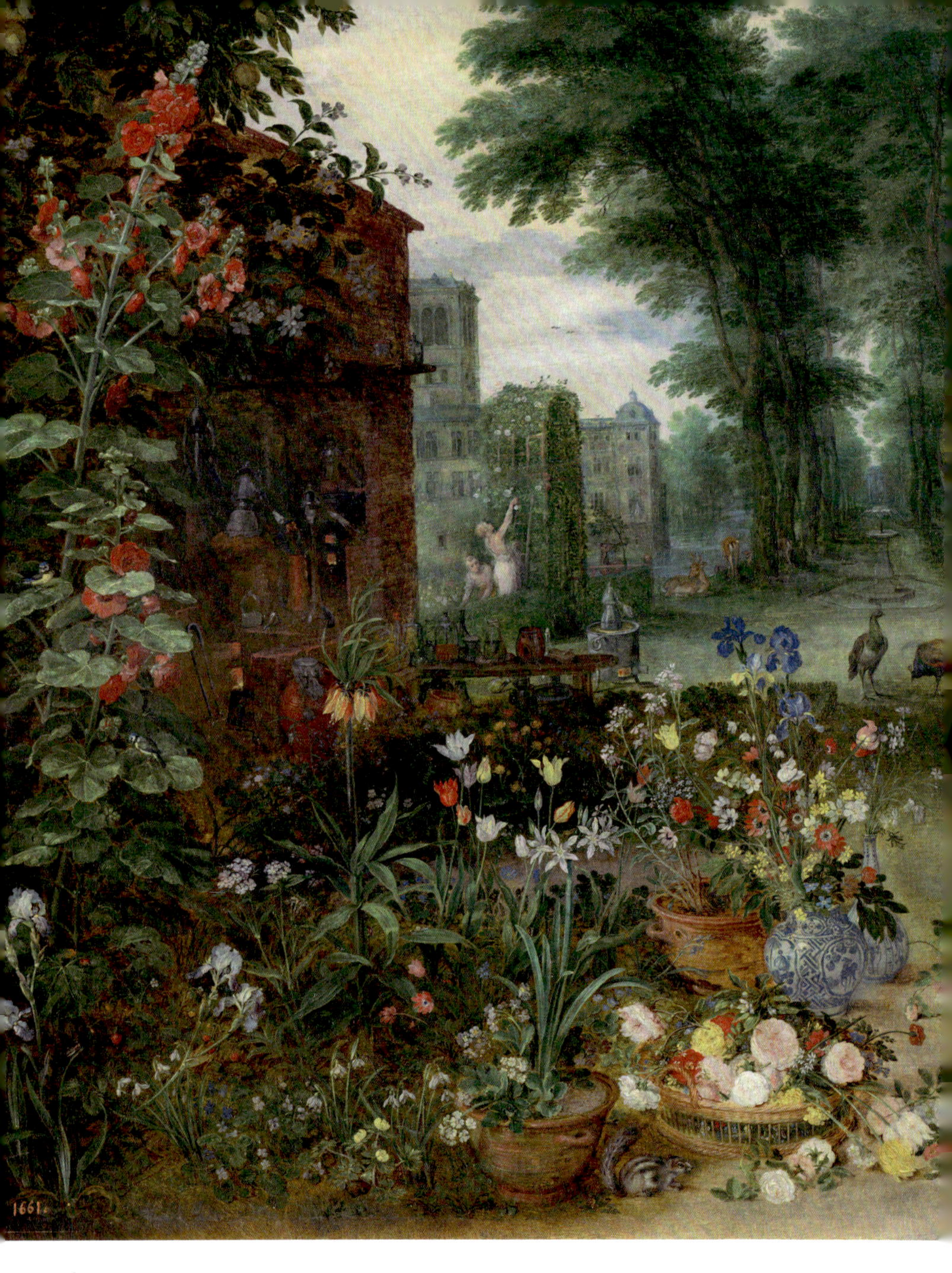
1661

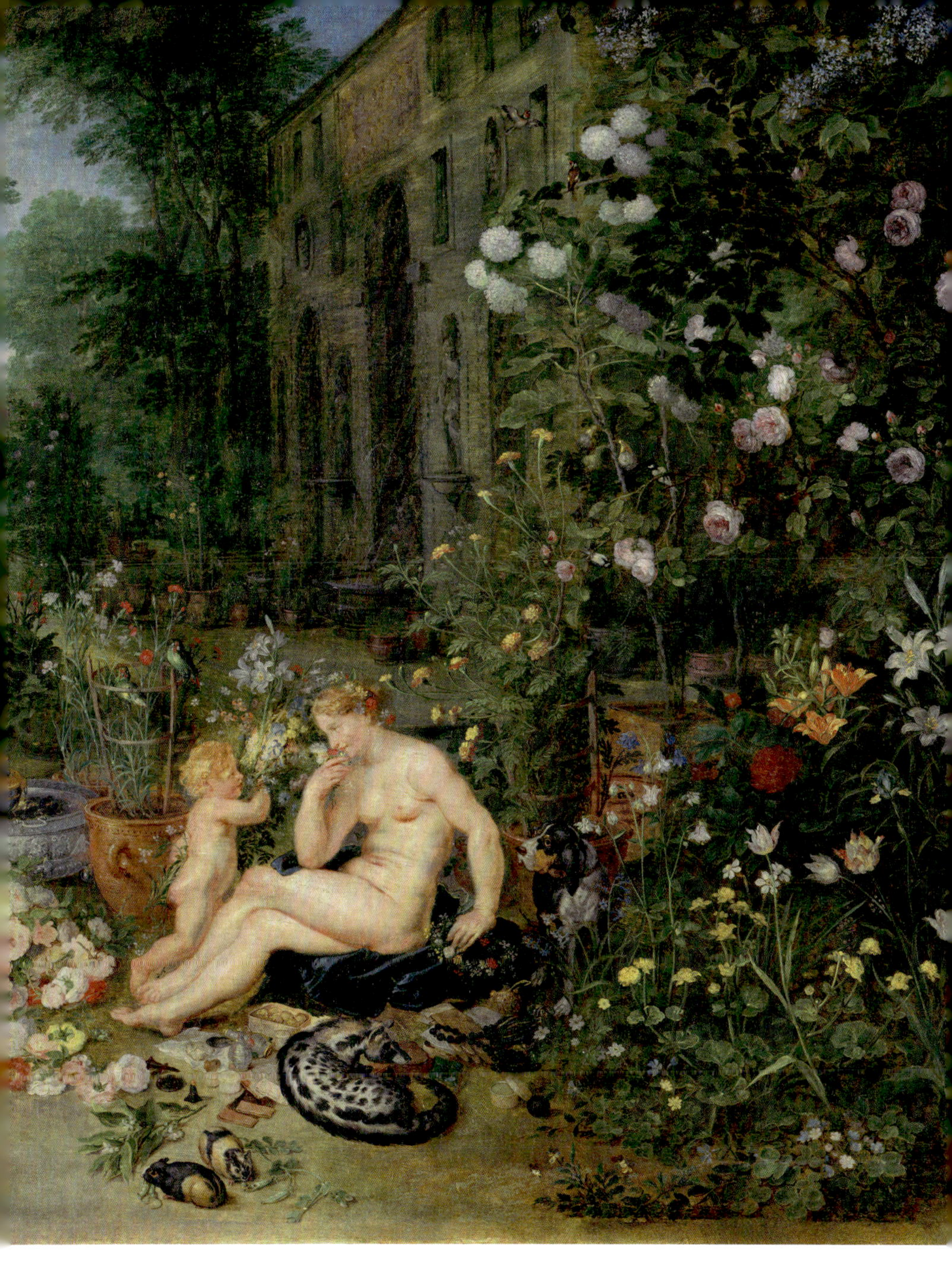

Jan Brueghel I and Peter Paul Rubens, *Allegory of Smell*, c. 1617, Museo nacional del Prado, Madrid

Le bouquetier [...] *sera disposé et sis* [sic] *à l'entrée principale du jardinage: à ce qu'aiant pour premier objet ès* [sic] *parterres, les beaux compartiments, le plaisir s'en rapporte plus grand à la veue, le rencontrant d'abord, que s'ils en estoient plus ésloignés. Ce n'est toutes-fois de l'avis d'aucuns, qui, avec raison, tiennent le bouquetier reculé et comme caché: à ce que vue le dernier, il soit estimé, à la manière que les marchands font admirer les fines estoffes, après avoir faict monstre des grossières.*

The flower garden... will be planted and situated at the main entrance to the garden so that it will be the first thing that people see in the parterres, in the most beautiful compartments. The pleasure will be even greater when you see it first than when it is further away. But not everyone shares this view. Some prefer to keep the flower garden at a distance and hidden, and for good reason. As such, the garden only reveals itself at the last moment, like a merchant showing off his finest fabrics after having first shown the coarser ones.

Olivier de Serres, *Le théâtre d'agriculture et mesnage des champs*, 1600

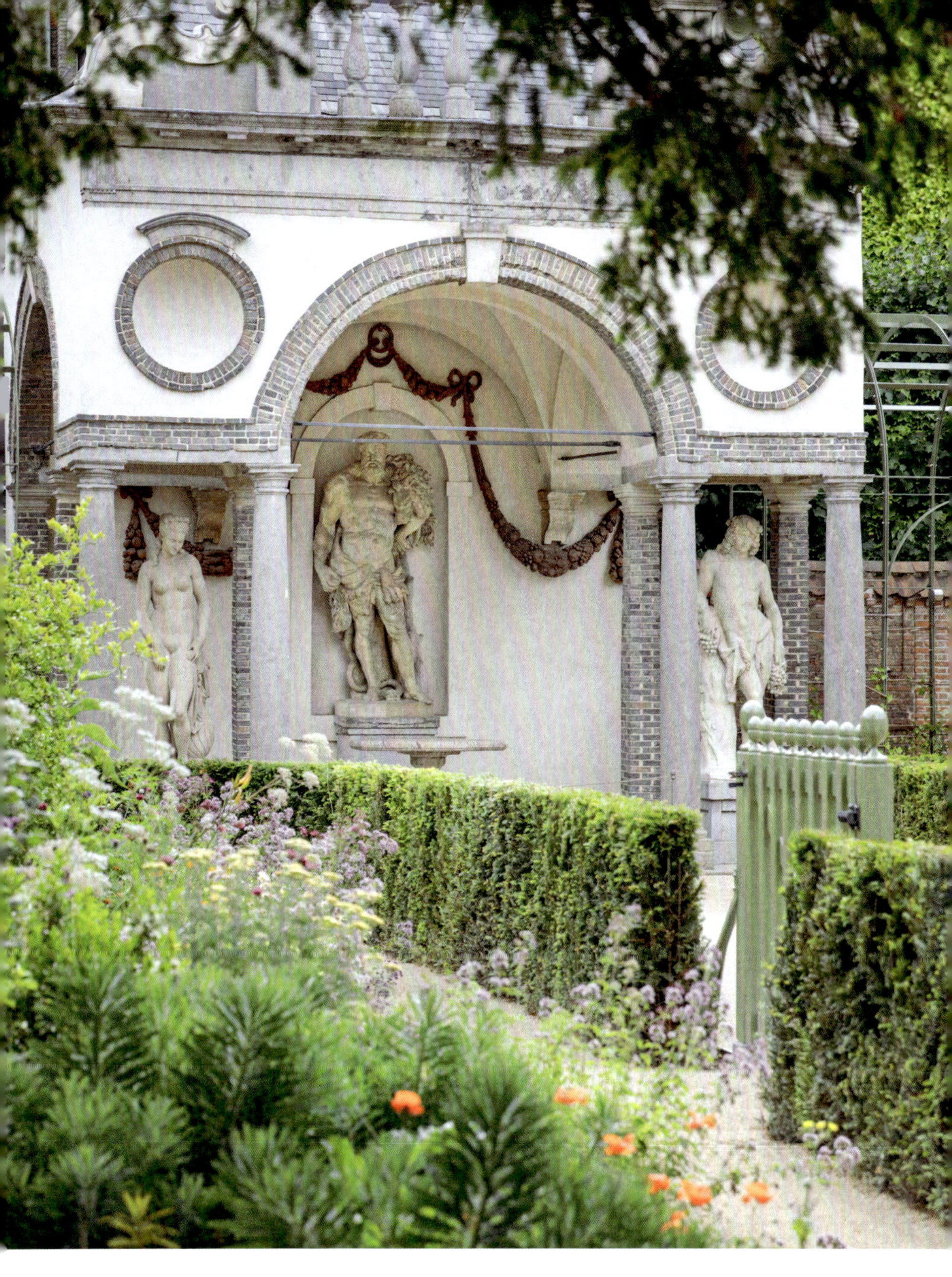

View from the parterre to the garden pavilion

Antwerp city gardens in Rubens's time

Ever since the evocation of the Rubens Garden created by Emiel Van Averbeke and Georges Wachtelaer between 1937 and 1946, the ambition has always been to use a selection of historical plants. However, the designers had no real historical research to rely on in those days. This was mainly because only a handful of documents that are directly related to Rubens's garden have survived. Thanks to a note from Rubens, the settlement of his estate, and a 'memorial' from his second wife Helena Fourment, we can be sure that figs, pears, oranges, and limes were growing here by the end of Rubens's life. The small *Florilegium* in this book discusses these species – their appearance, varieties, and the use people made of them as a cooking ingredient and for medicinal purposes – along with some other flowers and plants in further detail. In his biography of Rubens, Philip Rubens, the artist's nephew, described how his uncle bought a large house in Antwerp and remodelled it in the Roman style, and how he laid out a spacious garden with all kinds of trees (in Latin: *omnis generis arboribus*). Even though Philip wrote the biography after Rubens's death, the diversity of trees must have been a defining feature of the garden even in his day.

The new garden design required new historical research. For the first time in the museum's history, Antwerp city gardens from Rubens's time were examined. This information could not be found in a book or in one specific source or archive. Estate inventories – the preferred source for researching material culture in the early modern period – only mention a limited number of species of flowers (such as carnations and jasmine) and orangery plants (such as orange, bay tree, fig, pomegranate, oleander, lemon, and lime). In most instances, these kinds of documents list plants in tubs and pots. Flowers, trees, and shrubs that were planted in soil were not considered movable property, which is why the notary's clerk would not include them in the inventory. There are exceptions, however, such as the estate inventory of Alderman Charles de Tassis. In the spring of 1641 – one year after Rubens's death – his garden on the Stijfselrui behind the Capuchin monastery was in bloom, with auricles, anemones, lilies, liverworts, hyacinths, daffodils, crocuses, sowbread, carnations, fig trees, pomegranate trees, wild bay trees, and a myrtle tree, among others. Besides several gardening tools, De Tassis's

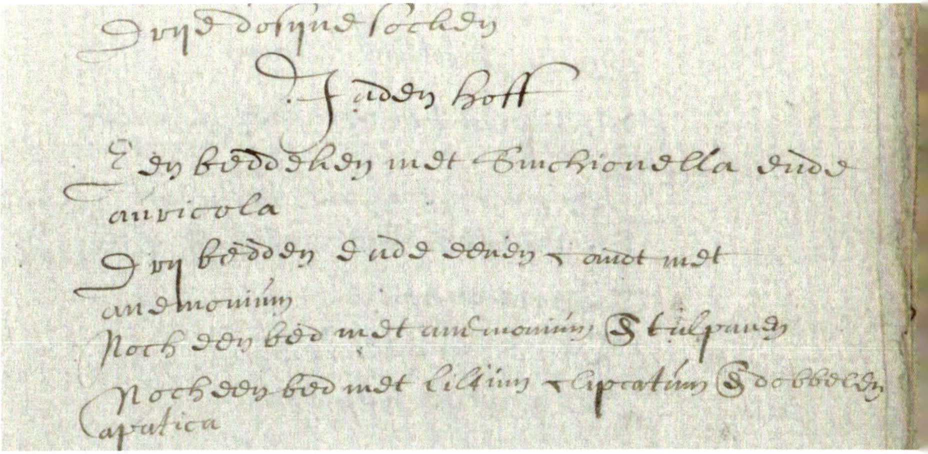

Extract from the estate inventory (1641) of Alderman Charles de Tassis, Felixarchief, Antwerp

estate inventory provides a meticulous description of the flower beds with their nearly 20 different species.

For the identification of plant names in 17th-century Antwerp city gardens, we had to delve deeper into all kinds of (sometimes with many lacunas) archival documents, such as rental and sales inventories, (loose) accounts, journals, wills, litigation files, letters, notes, and travel journals. For the new Rubens Garden, more than 50 documents yielded a particularly rich and colourful plant palette of almost three hundred flowers and plants in Antwerp gardens from the first half of the seventeenth century. The species that are most frequently mentioned are tulips, bay, orange, cherry, pomegranate, figs, grapes, limes, carnations, roses, apricots, daffodils, plums, apples, currants, and lemons. As part of the garden's redevelopment, the designers chose plants that were referenced in historical source material from Rubens's time where possible.

Allons voir en Avril, les Iardins de Bruxelles, ou beaucoup de Seigneurs, beaucoup de Damoiselles, font parade à monstrer les singularitez d'un grand nombre de fleurs & leurs diversitez, tout le mesme en Anvers, en trafique opulente, superbe en ses maisons & magnifique en plante [...].

In April, let us go and see the gardens of Brussels, where many lords and noblewomen show off the peculiarities of a large number of very different flowers, like in Antwerp, with its flourishing trade, superb houses, and magnificent plantings....

Jean Franeau, *Jardin d'hyver ou cabinet des fleurs*, 1616

From the library of Rubens and his son Albert

Garden owners like to draw on horticultural literature for inspiration, tips, and advice. Rubens and his eldest son Albert were no different in this respect. On 2 February 1615, the artist bought the *Hortus Eystettenis* in the Antwerp bookshop of Balthasar I and Jan II Moretus. This new and very richly illustrated florilegium discussed Prince-Bishop Johann Conrad von Gemmingen's herb and ornamental garden around his castle in Eichstätt. The 367 drawings and engravings of a total of 1,084 ornamental, medicinal, and edible plants were done by various artists under the supervision of the pharmacist and botanist Basilius Besler.

The *Hortus Eystettensis* was printed at Nuremberg in 1613, in a very exclusive edition of three hundred copies. Two editions exist, both printed in large format (57 × 46 cm): a black-and-white edition with text, and a deluxe edition with coloured illustrations on high-quality paper without text. All three hundred books were sold in four years. The original drawings that preceded the engravings were based on carefully observed sketches that were drawn to life. They are kept at the University of Erlangen.

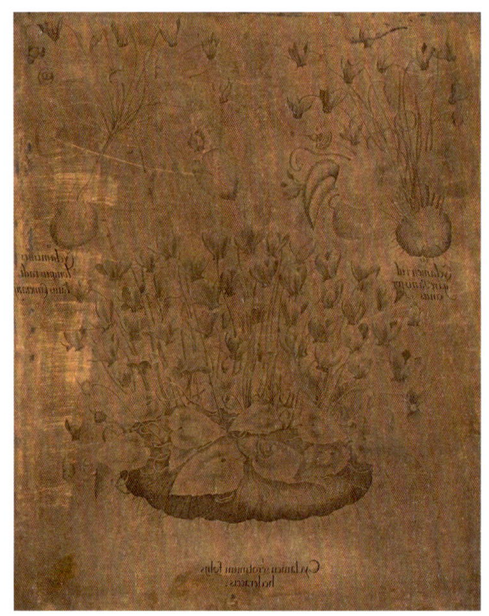

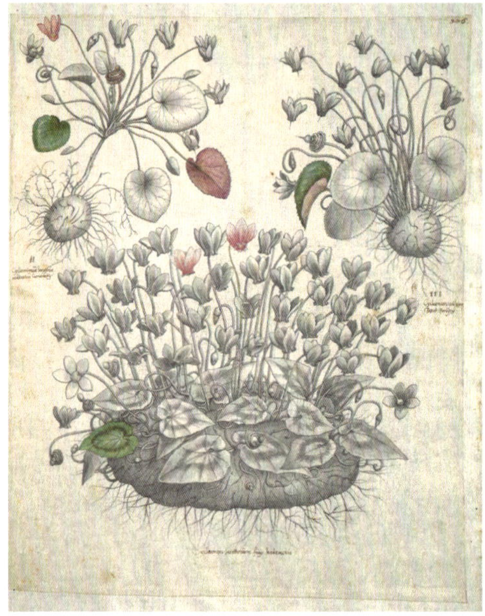

Copper plate and partly coloured drawing of *Cyclamen hederifolium* in preparation for the engraving in the *Hortus Eystettenis*, Albertina, Vienna & Universitätsbibliothek, Erlangen-Nürnberg

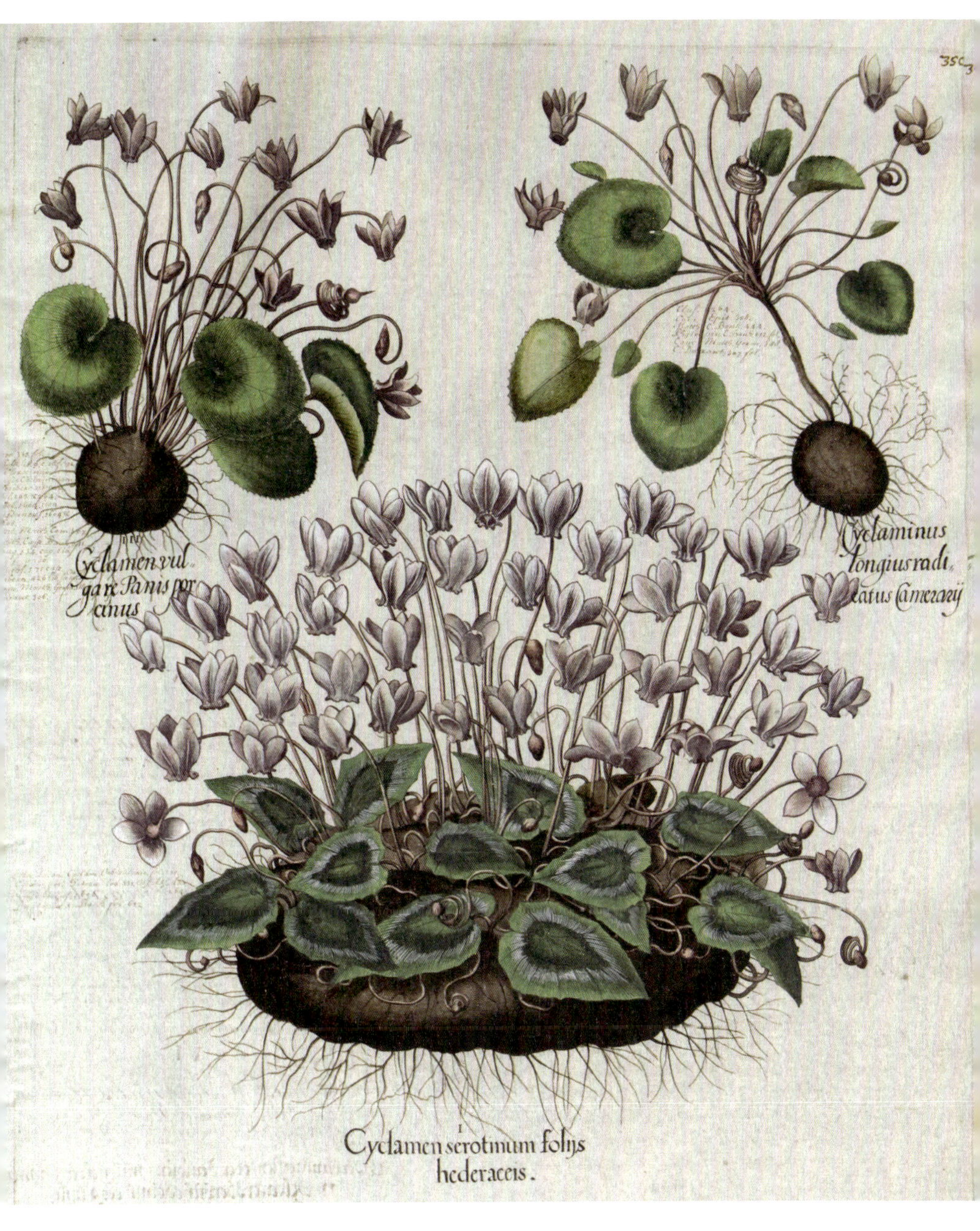

Engraving of *Cyclamen hederifolium* in a coloured edition of the *Hortus Eystettenis*, 1613, Teylers Museum, Haarlem

Although Rubens's collection of books became dispersed after his death, we have a good idea which books would have lined his library shelves. Rubens paid 98 guilders to Moretus for his copy of the *Hortus Eystettensis*, possibly making it the largest and most expensive book in his collection. We cannot be certain whether he owned a coloured copy or a black-and-white edition. Uncoloured, unbound copies started at 35 guilders. A coloured luxury version would likely cost five times more. We can be certain, however, that very few other Antwerp citizens – perhaps nobody – owned a copy of Besler's book. Besler's *Florilegium* served as inspiration for the new museum garden's plant palette, but its influence as source material for Rubens's garden should not be overestimated. The work takes you through a unique, early 17th-century garden with many rare plants, which in those days would have been shared by a small network of botanists and collectors. Rubens's garden was definitely not a copy of this exceptional garden.

Among the many books on architectural theory in Rubens's library, there were also two works by the engineer Salomon de Caus, including *Les raisons des forces mouvantes*, in which all kinds of designs for caves and fountains with surprise effects were developed, in text and visuals. After Rubens's death, the *Hortus Eystettensis* and the works of De Caus ended up in the bookcase of Rubens's eldest son Albert. This also contained several important botanical works: the *Historia Stirpium Commentarii Insignes* by Leonhart Fuchs and the *Kruydtboeck* by Matthias Lobelius. Lobelius's book was published by Christopher Plantin in 1581 and featured more than 2,000 woodcuts of plants. Albert also owned a copy of *L'agriculture et maison rustique* by Charles Estienne

(1564), a gardening and farming manual. There is a good chance that these books originally came from Rubens's library, as Albert inherited all his father's books. But Albert was also an avid book collector, and the collection of books of Jan Brant, his maternal grandfather, had also been left to him.

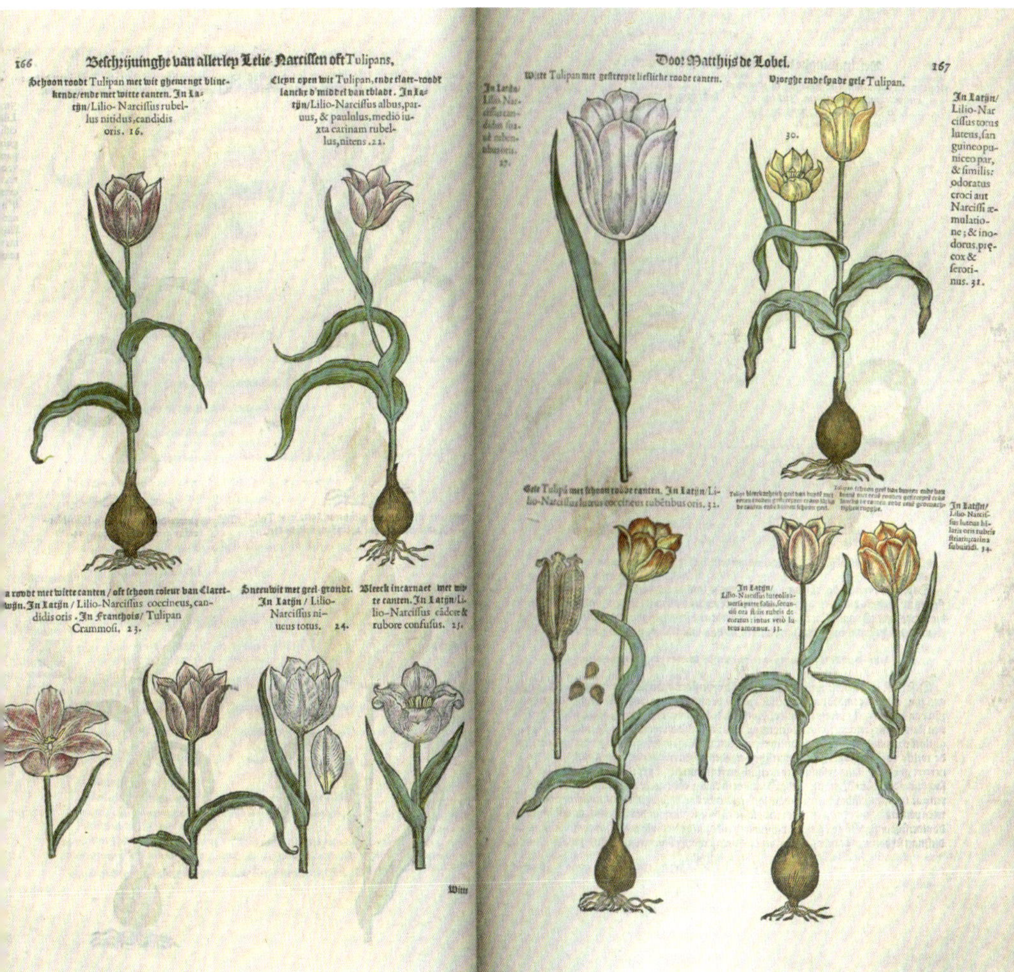

Matthias Lobelius, *Tulips* in the *Kruydtboeck*, 1581, Museum Plantin-Moretus, Antwerp

The flower catalogue of a Brussels painter

Charles of Arenberg and his wife Anna of Croÿ-Chimay created *'un fort beau et gran jardin'* (a large and very lovely garden) at Enghien from 1607. In the winter of 1609, the Italian botanist and flower supplier Matteo Caccini asked Charles of Arenberg for a catalogue of his garden. As the garden was not yet ready, and in anticipation of the plant catalogue of his own garden, Arenberg sent Caccini an alphabetical catalogue of flowers in Dutch by a Brussels painter friend of his, *'uno dely piu grande harbarista (sic) et meglio provisto'* (one of the greatest and best-equipped herbalists). This inventory of the Brussels painter's garden offers a particularly detailed insight into an early Baroque spring garden. The names of hundreds of tuberous and bulbous plants, including anemones, crocuses, hyacinths, daffodils, peonies, and tulips in all kinds of stunning colours which, according to the catalogue, 'are impossible to define and name', served as a source for the planting plan of the new Rubens Garden. Illustrations from *Hortus Floridus*, Crispijn de Passe II's popular florilegium from 1614, give us a good idea of what these flowers would have looked like.

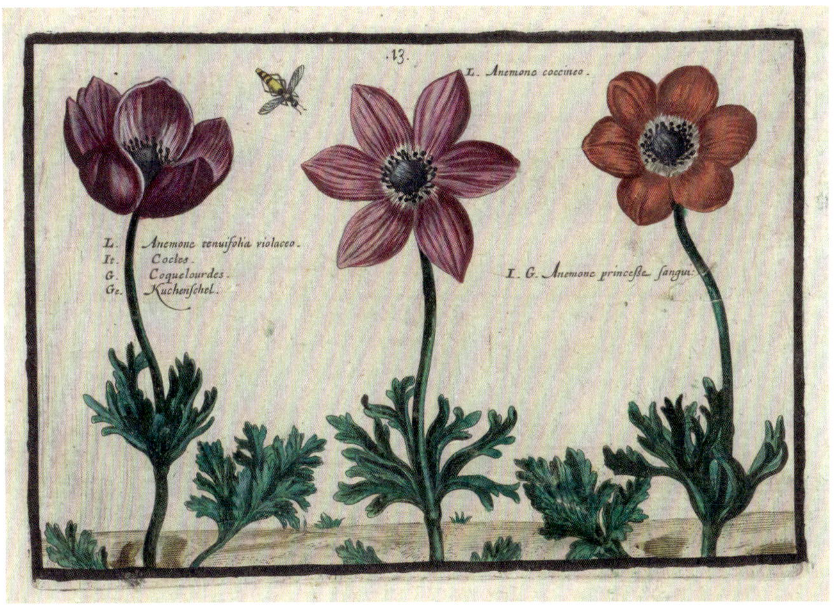

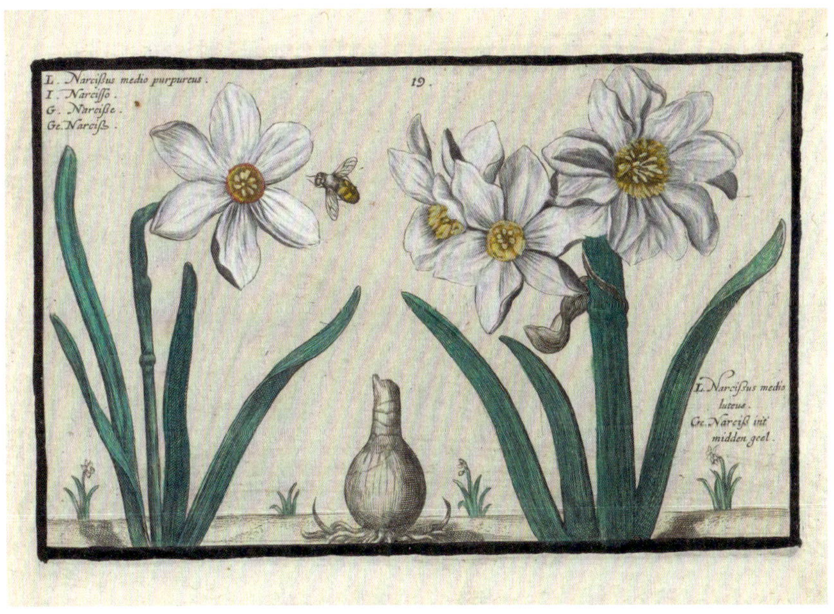

Crispijn de Passe II, *Narcissus* and *Anemone* from the *Hortus Floridus*, 1614, Oak Spring Garden Foundation, Upperville, Virginia (USA)

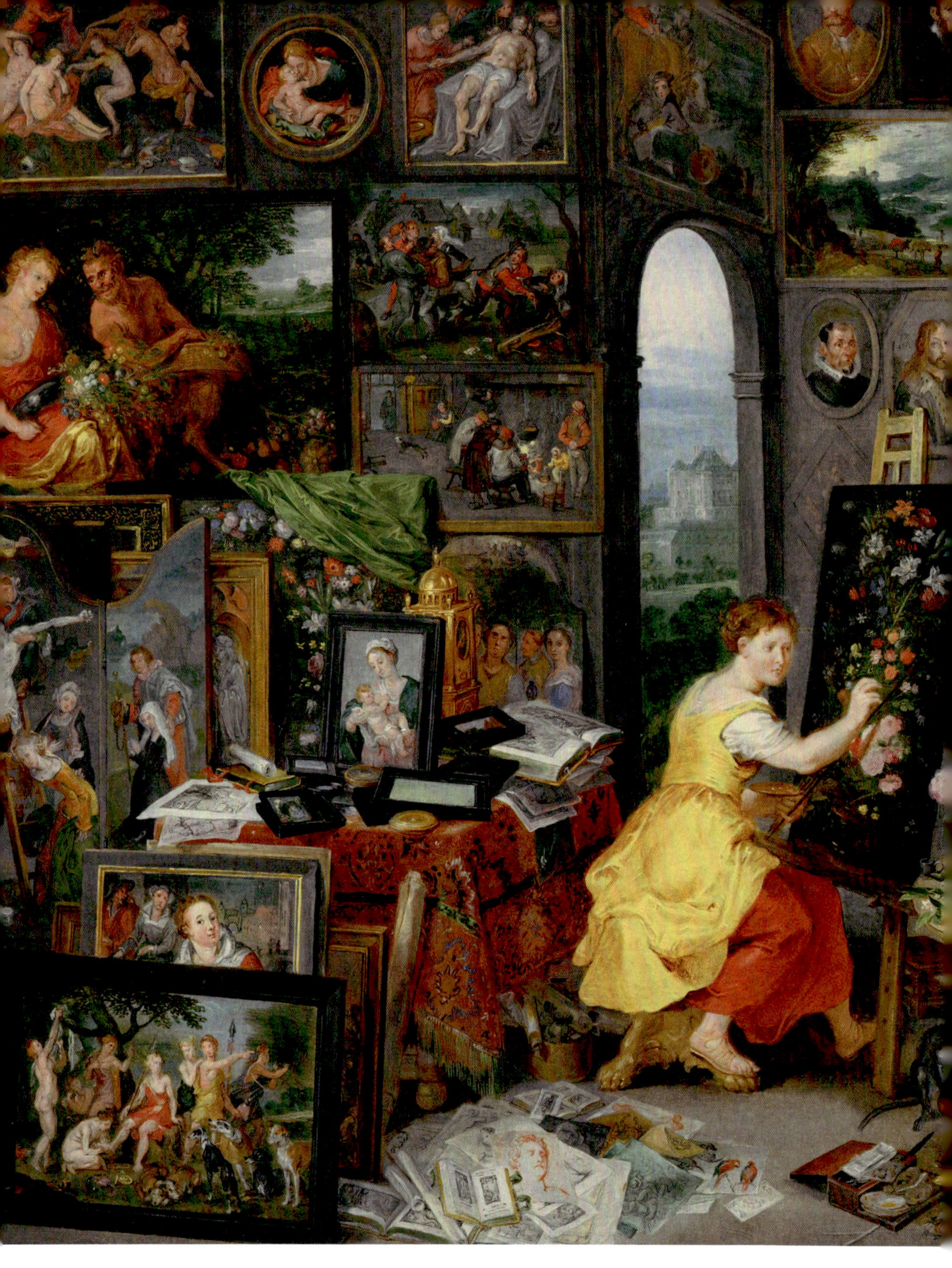

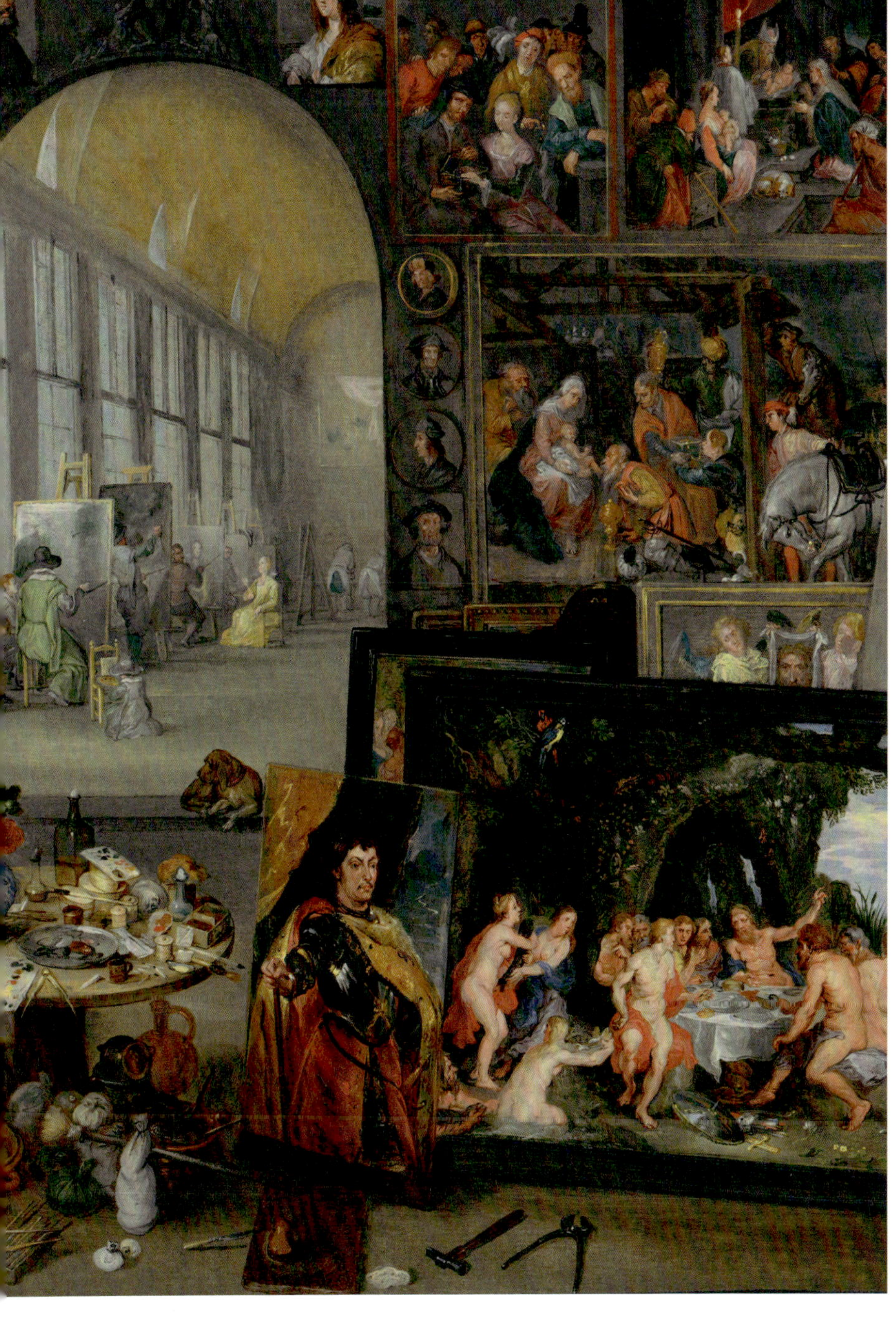

Jan Brueghel II, *Allegory of Painting*, c. 1625–1630, JK Art Foundation

Painted flower pieces for the Rubens Garden

Painted flower pieces also served as a resource for the planting plan of the new garden of the Rubenshuis. In a letter of 14 April 1606 to his patron Federico Borromeo, cardinal and archbishop of Milan, the Antwerp flower painter Jan Brueghel I emphasised the '*naturalleza*' (lifelikeness), the '*bellezza*' (beauty), and the '*rarita*' (rarity) of the flowers in the painting he had just started working on. In the spring of 1606, Brueghel did not yet have access to a garden with the rarest species in Antwerp, he wrote to Borromeo. To paint some flowers from life, the artist had therefore travelled to Brussels, where he may have visited the garden of the fellow painter, whose aforementioned 1609 flower catalogue survives. On 25 August 1606, he informed Borromeo that he had finished the painting: 'I don't believe that anyone has ever painted so many rare and different flowers together, or finished them with such precision. It will be a stunning sight to behold in winter. Some colours are only slightly different from the natural ones'.

The correspondence between Brueghel and Borromeo shows that there was a direct link in the early 17th century in the Southern Netherlands between flowers in gardens and their representation

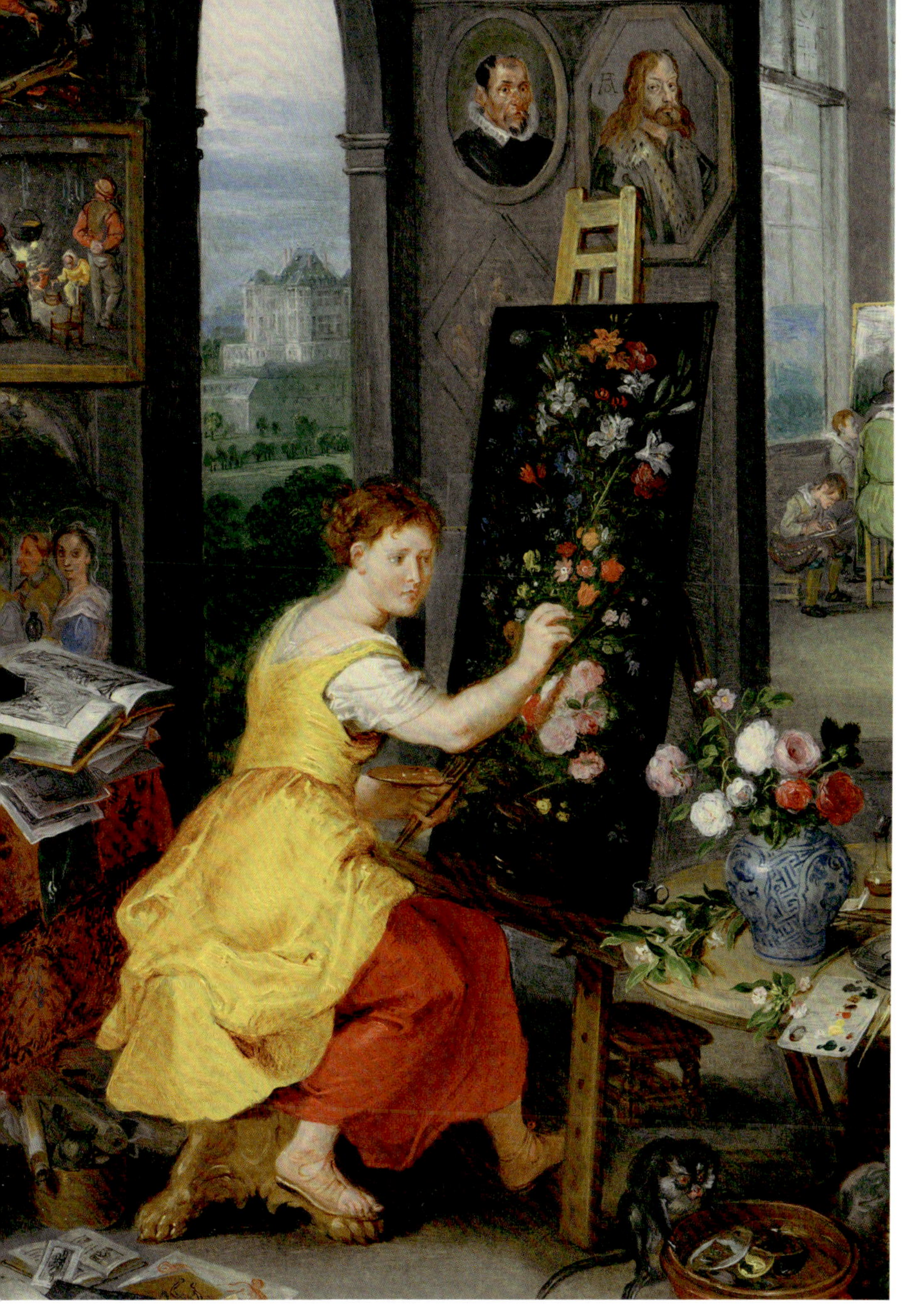

DETAIL FROM Jan Brueghel II, *Allegory of Painting*, c. 1625–1630, JK Art Foundation

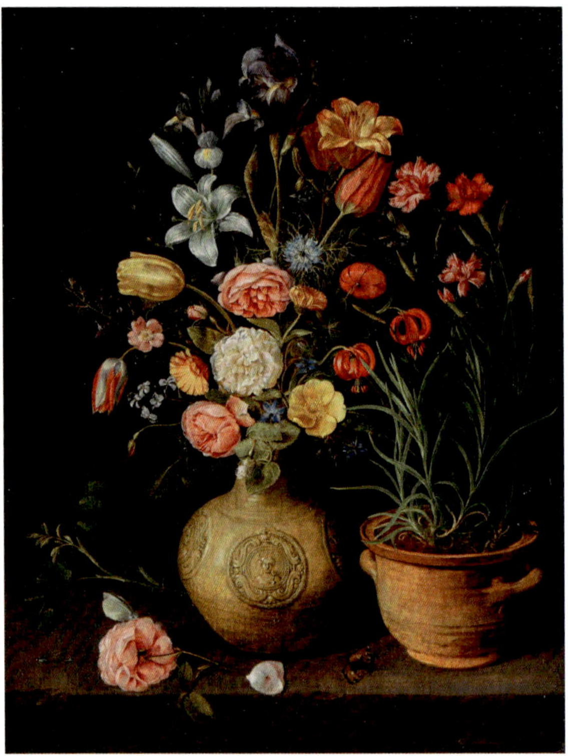

in a painted flower piece. Brueghel and many of his fellow artists painted flowers from different seasons with such precision that enthusiasts and collectors were able to recognise them in a painted flower still life. Even today we can still name the many species in still lifes – relying on our knowledge of plant books and florilegia of that period. We can now pick a bouquet for the Rubens Garden from what was painted then, with great care and meticulous attention.

Clara Peeters, *Flowers in a Stoneware Vase, with a Pot with Carnations*, 1612, private collection

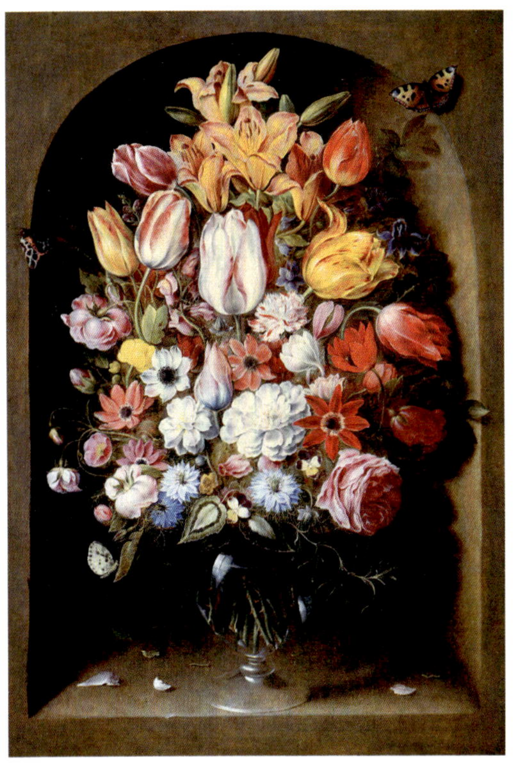

For the new design of the Rubens Garden, all species in 15 flower pieces from the period between 1605 and 1660 were carefully determined by still life specialist and biologist Sam Segal. Moreover, these flower still lifes were created by painters who collaborated with Rubens, such as Jan Brueghel I and Osias Beert. We also looked at flowers that were painted by Antwerp-based contemporaries of Rubens, such as Andries Snellinck and Clara Peeters, and works by painters whose work Rubens collected. By the end of his life, he owned at least eight Antwerp flower pieces and other still lifes with flowers: six were by Frans Ykens, one by Frans Snyders, and one by Daniël Seghers.

Osias Beert, *Flowers in a Glass Vase with a Niche*, 1610–1620, Snijders& Rockoxhuis, Antwerp

Sometimes these flower pieces consisted of just one species. In 1636, Alexander Adriaenssen painted a vase containing only tulips in various stages of bloom. Ten years later, he revisited the concept, but with roses: from bud to full flower and even in their spent state. Antwerp flower painter Philips de Marlier painted a vase containing only carnations. In 1612, Clara Peeters combined a terracotta pot with carnations and a lush bouquet in a vase, creating a unique still life. In the research for the Rubens Garden, Jan Brueghel I is represented by a painted flower piece of a wooden tub with 130 species and varieties of flowers, the largest and most extraordinary number to have yet been found in a floral arrangement.

The research on flower still lifes shows that there was a wide range of varieties of bulbous plants and species with rootstalks available in 1600–1620. Tubers, such as sowbread, often appear at the bottom of the bouquet. Artists also painted fragrant herbs, such as rosemary. Small flowers, such as liverwort and sweet violet, were popular. Other frequently featured species include love-in-a-mist, snake's head fritillary, lily of the valley, various irises, daffodils and crocuses, marsh marigold, striped anemone, apothecary's rose (*Rosa gallica*), damask rose (*Rosa damascena*), eglantine rose or sweet briar, French marigold, wild pansy, annulated sowbread (cyclamen), borage, cotton lavender, and crown imperial.

In general, the number of species in flower still lifes that were painted between 1600 and 1620 is more limited than in later periods. In the following decades, emphasis was placed not on the number of species but on the many variants per species, such as tulips in particular, but also daffodils, hyacinths, and irises. Roses also became increasingly popular. Relatively common species include poppy-flowered anemone, carnation, giant larkspur,

RIGHT Jan Brueghel I, *Flowers in a Wooden Tub*, c. 1607–1608, Kunsthistorisches Museum, Vienna

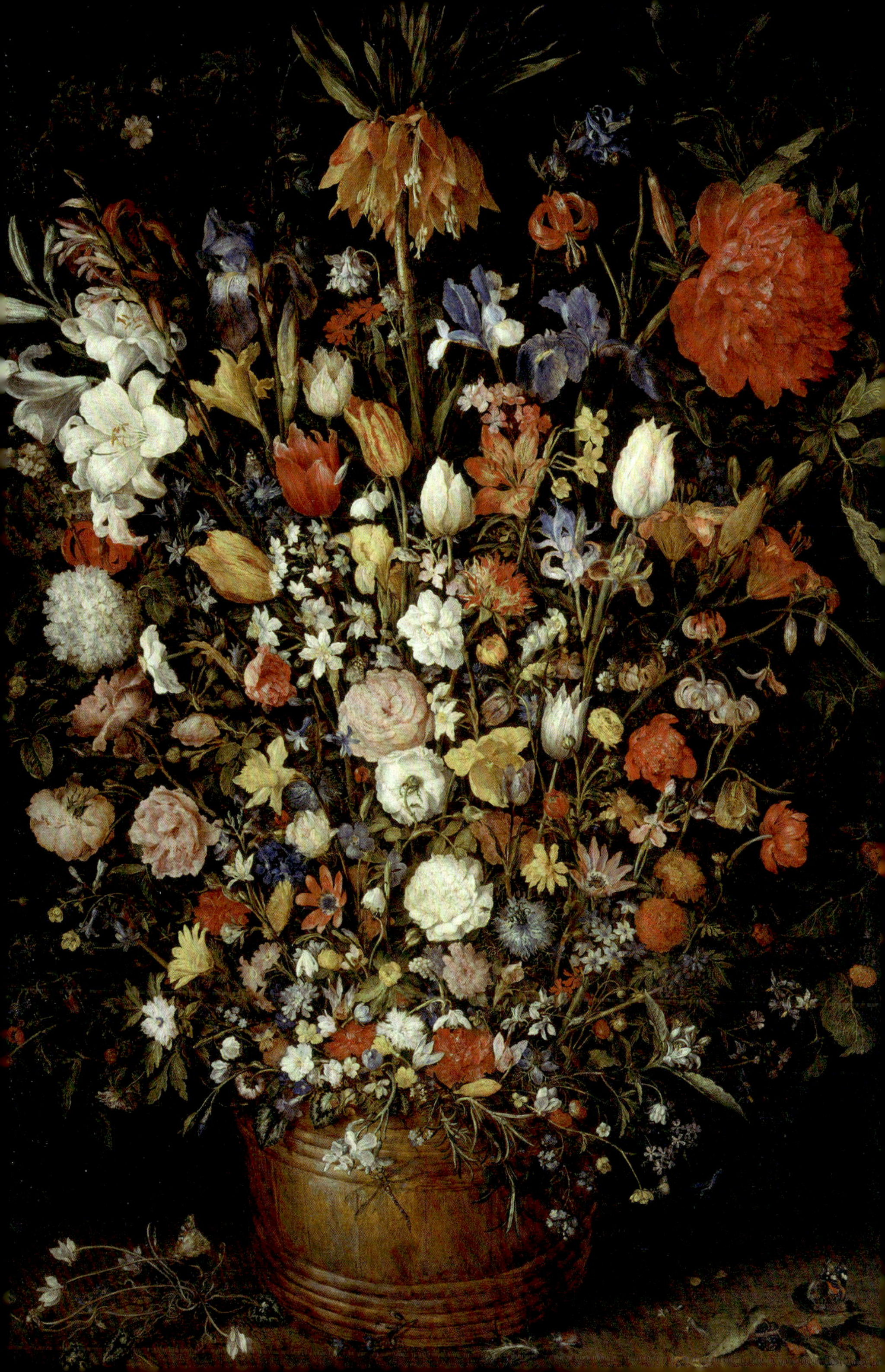

Maltese or Jerusalem cross, saxifraga or rockfoils, field gladiolus, marvel of Peru, dark columbine, and hawthorn.

In Rubens's time, flowers were not just cherished for their beauty and rarity. Depending on the context and the spectator's knowledge, they could also take on a symbolic meaning. In allegorical representations, flowers symbolise earth as an element, spring as a season, or scent as a sense, as in the *Allegory of Smell* that Jan Brueghel I and Rubens painted around 1617. As the middle of the five senses, between intellectual sight and hearing and physical touch and taste, flowers were seen as the connection between body and spirit, between the earthly and heavenly kingdoms.

Flowers, in general, symbolised transience, but in the 17th century, people also attributed specific symbolic properties to individual species. Roses, for example, have been associated with love since ancient times. Small or low-growing plants, such as daisies and sweet violets, were often associated with humility, much like hanging flowers such as lily of the valley and snake's head fritillary. Upward-facing flowers, such as tulips, reveal their dependence on the sun or sky, especially if they turn with the sun during the day, like marigolds or sunflowers. Plants with spines or thorns could be associated with the suffering of Christ and martyrs. Climbing plants, such as ivy and grapes, symbolised connection and friendship. The number three in the leaf shape of clover species, aquilegia, strawberry, liverwort, peony, ivy and grapes, and in the colours of the tricoloured violet represents the Holy Trinity. Fragrant and sweet flowers, such as jasmine and orange blossom, were associated with virtues, the Virgin Mary's virtues in particular, whereas bitter-tasting plants were equated with suffering. Some flowers are beautiful but malodorous, such as crown imperials, which can symbolise vanity, among other things. Evergreens such as ivy, citrus and the bay tree were seen as symbols of eternity or everlasting glory. Trees, in general, often symbolise life force. Fruit stands for fertility, flavour or autumn, grains for summer, and edible roots and tubers for winter. Spring plants, such as winter aconite, liverwort and crocuses, were associated with resurrection and the hope of salvation in a religious context.

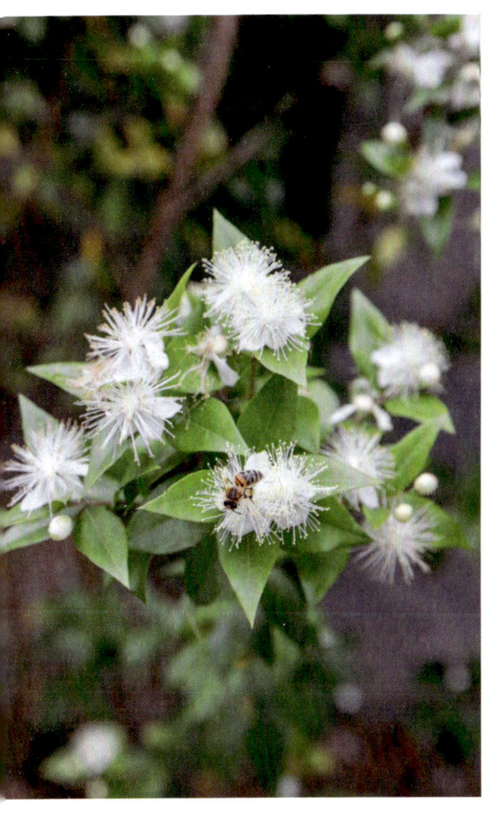

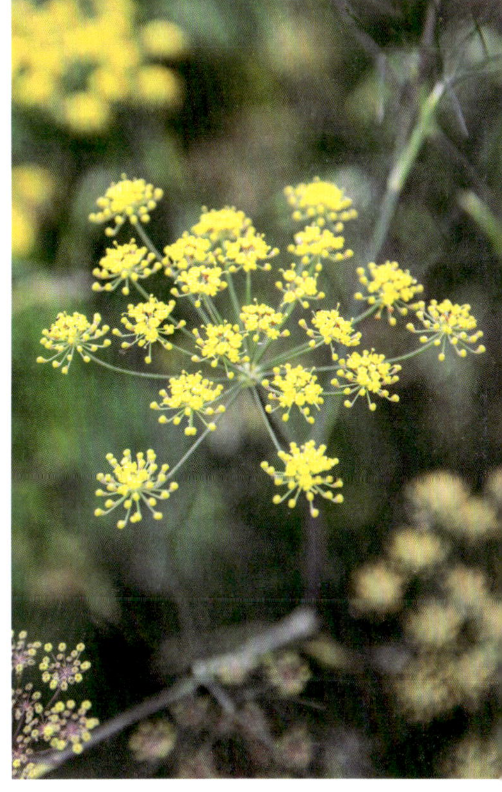

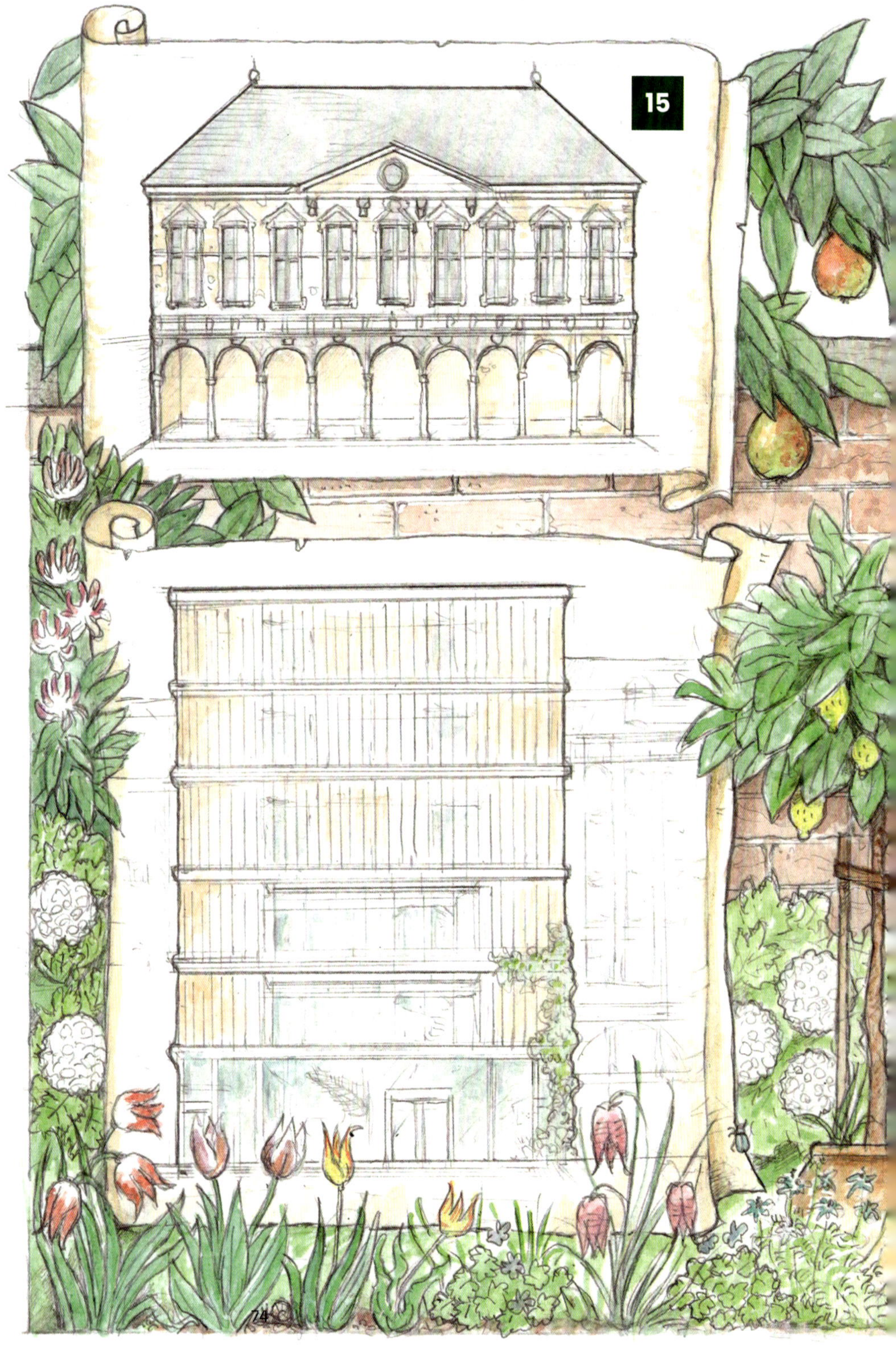
15

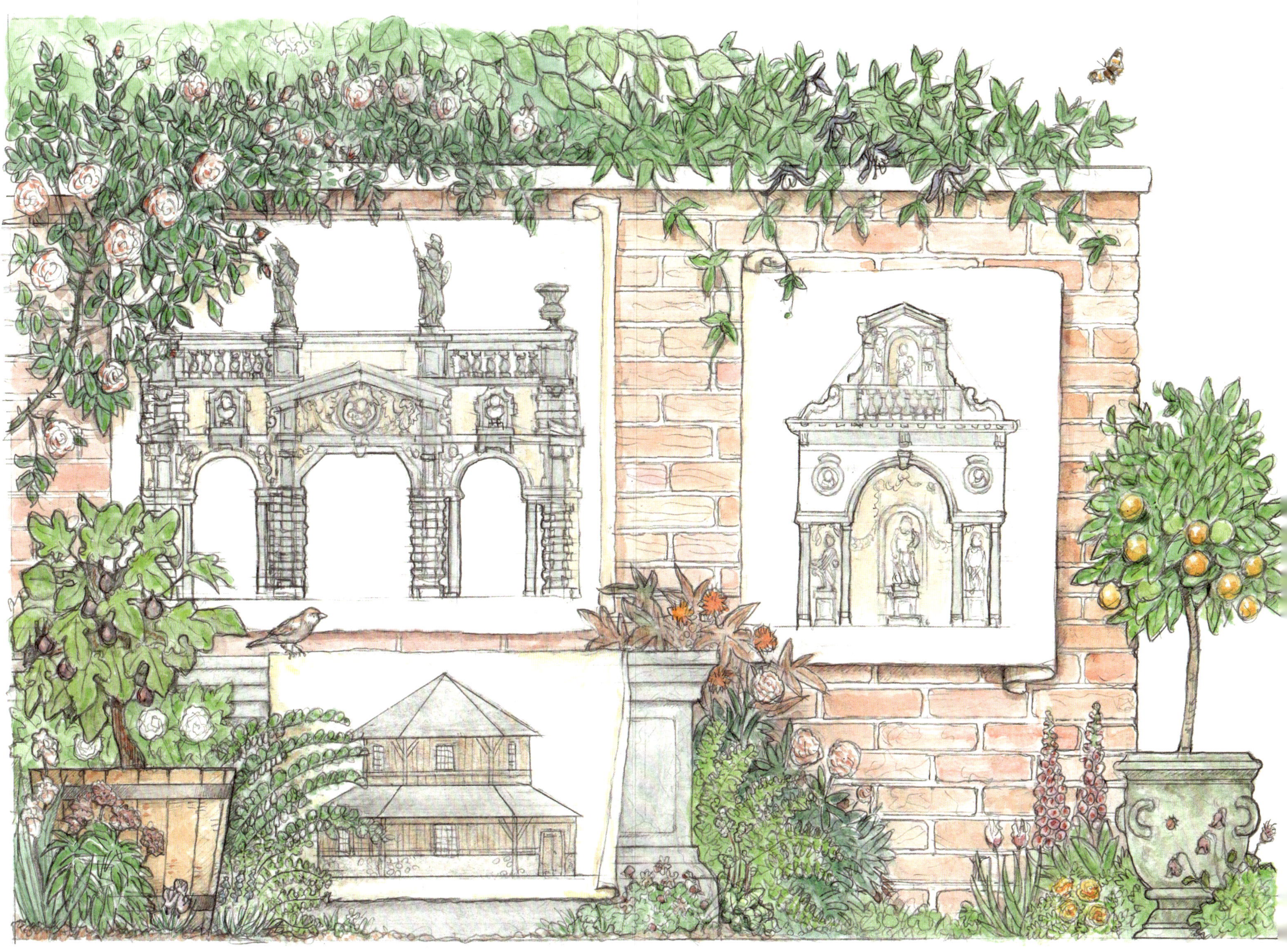

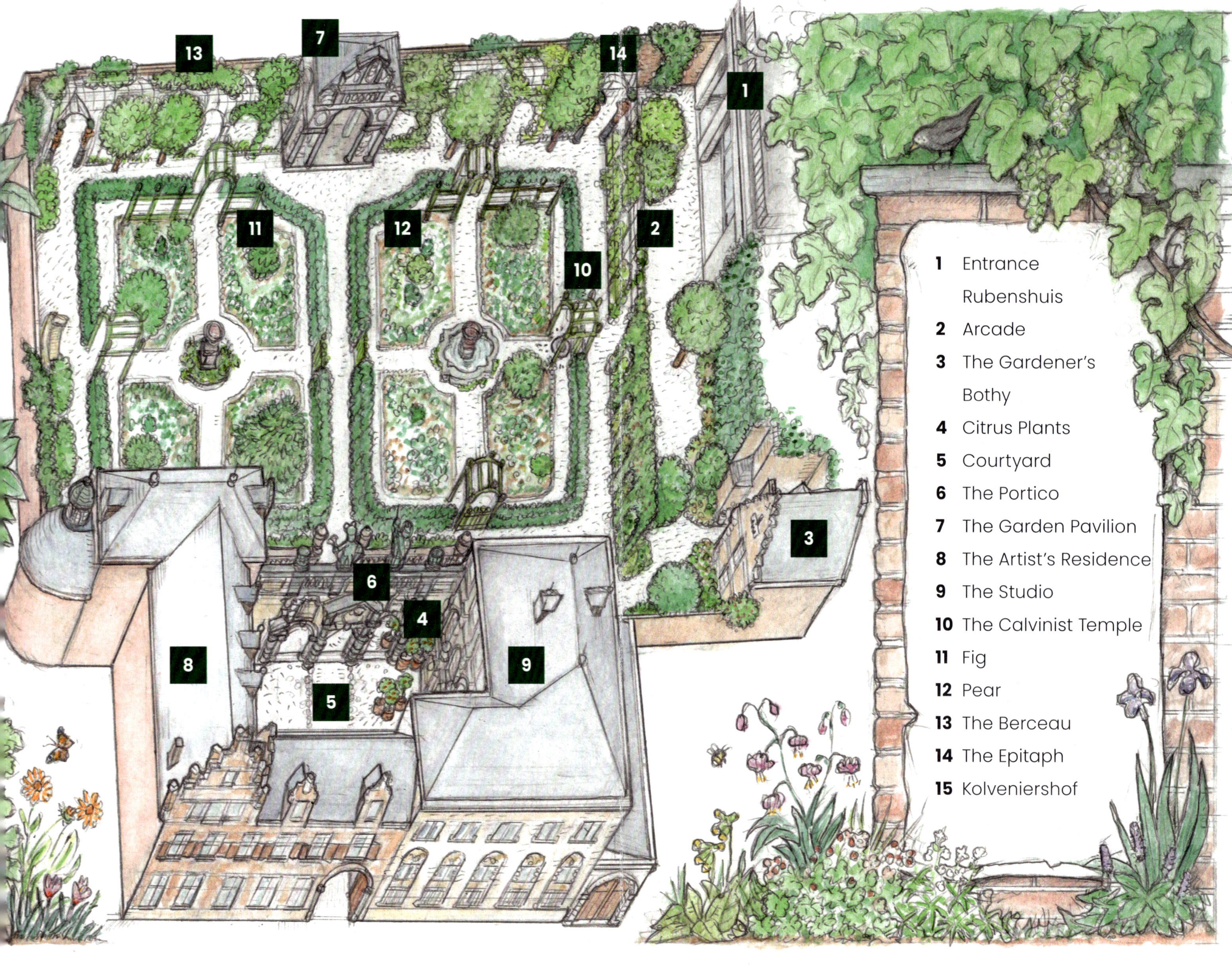
13
7
14
1
11
12
2
10
3
6
4
8
9
5
1 Entrance Rubenshuis
2 Arcade
3 The Gardener's Bothy
4 Citrus Plants
5 Courtyard
6 The Portico
7 The Garden Pavilion
8 The Artist's Residence
9 The Studio
10 The Calvinist Temple
11 Fig
12 Pear
13 The Berceau
14 The Epitaph
15 Kolveniershof

A SMALL FLORILEGIUM OF THE RUBENS GARDEN

Aquilegia vulgaris

Columbine

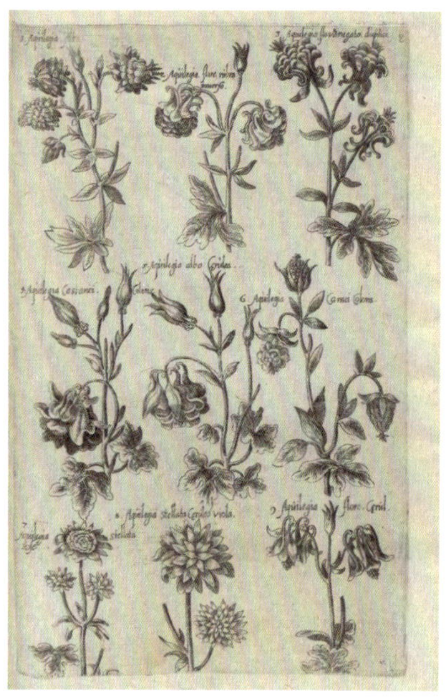

The graceful ***Aquilegia vulgaris*** or common columbine flowered in Rubens's garden in early summer. In ***Den kleynen herbarius ofte kruydt boecxken*** (Antwerp, 1640), columbines are described as 'very pretty flowers'. These elegant perennials were therefore often included in 17th-century flower still lifes. The seed of aquilegia was used as a cooking ingredient and against certain ailments. Aquilegia were praised for the diversity of their colour and form in the botanical works and gardening manuals of that period. 'For one can think of almost no colour that this flower does not come in, except perhaps yellow', according to Rembert Dodoens in his ***Cruydt-boeck*** of 1644.

Aquilegia vulgaris in the *Florilegium* by Emanuel Sweerts, 1614, Hendrik Conscience Heritage Library, Antwerp

According to Lobelius, this plant thrived in cool southern climes and regions. In France and England, they were found in meadows, 'but the flowers are neither as beautiful nor as bright as those that grow in gardens: for they have two, three or four doubled flowers and thicker and brighter stems, with flowers in many colours, or each flower in one colour, or all three colours in one flower: namely white, purple and blue'. According to Lobelius, the flower was also perfect for flower posies 'for its flowers are a very nice purplish-blue, made of hollow spurs, with a shape that is very similar to the neck and beak of a dove, which is why it is also called columbine in English'. *Den verstandighen hovenier over de twaelf maenden van 't jaer* – the Flemish version of the gardeners' handbook that was published in Amsterdam in 1669 and was printed in 1672 by Reynier Sleghers on Kammenstraat in Antwerp – describes some of the differences in these bell-shaped flowers: 'they can have single or double flowers, red and white, blue and speckled, like hanging bells, for your pleasure in the garden'. A motley collection of aquilegia is depicted in Besler's *Hortus Eystettensis* of 1613, of which Rubens owned a copy, but also in other (trade) catalogues, including Emanuel Sweerts's *Florilegium* of 1614.

Den kleynen herbarius of 1640 describes how 'when these flowers have faded, they produce black seeds in elongated seed pods'. Columbine seed was sometimes added to sugar or candied for use as confectionery sugar in those days. Lobelius described how the seed was widely used 'to get rid of spots and freckles in one's face'. According to Dodoens, columbine seeds could help against congested gallbladder and liver and also remediate dizziness and fainting.

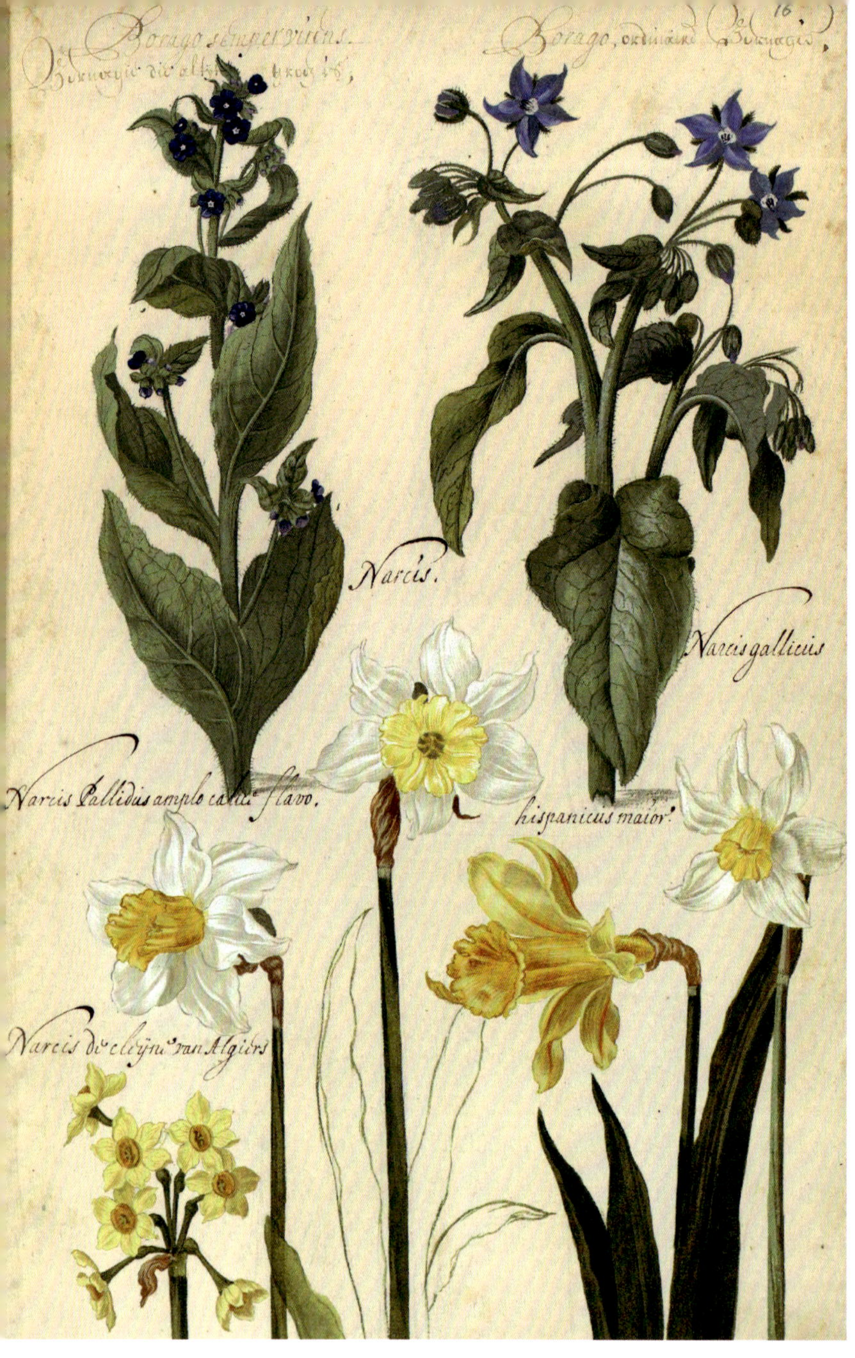

Franciscus de Geest, *Borage* in the *Hortus Amoenissimus*, 1668, Biblioteca Nazionale Centrale, Rome

Borago officinalis
Borage

In summer, the bulbs in the beds of the garden of the Rubenshuis make way for annuals, including borage. The striking flowers of this plant, which is native to Europe and the Mediterranean, were described by Lobelius in 1581 as 'a star-shaped bloom with five very pretty purple-blue and some-times white petals, with a small brown sepal'. The plant is edible and was said to lighten the mood.

According to the 1672 edition of *Den verstandighen hovenier over de twaelf maenden van 't jaer* as published in Antwerp, borage has broad leaves and 'beautiful star-shaped flowers that are bright blue in colour'. Both the leaves and the flowers are edible and were considered a natural mood booster. Dodoens has the following to say about borage in his *Cruydt-boeck* of 1644: 'These days, people don't just consume the herb (meaning its leaves) but also the flowers in wine. And they are added to salads with other herbs. And to boost happiness and merriment, to lighten people's mood'. The purple-blue flowers were also pickled with sugar and used in preserves, 'to strengthen the heart and banish all sadness and despondence and make people merrier'.

Borage blooms from early summer until autumn. Once sown, the plant often self-seeds in the same place in subsequent years. In 1668, Franciscus de Geest drew both the common and hardy varieties of borage (now *Pentaglottis sempervirens*) along with daffodils 'to nature' in the *Hortus Amoenissimus*. This stunning flower book contains 200 drawings of flowers from gardens in Leeuwarden, including many tulips. Jan van Kessel of Antwerp painted a sprig of borage along with a creeping thistle, butterflies, and other insects on an oak panel in 1654.

RIGHT Jan van Kessel I, *Insects with Creeping Thistle and Borage*, 1654, The National Gallery, London

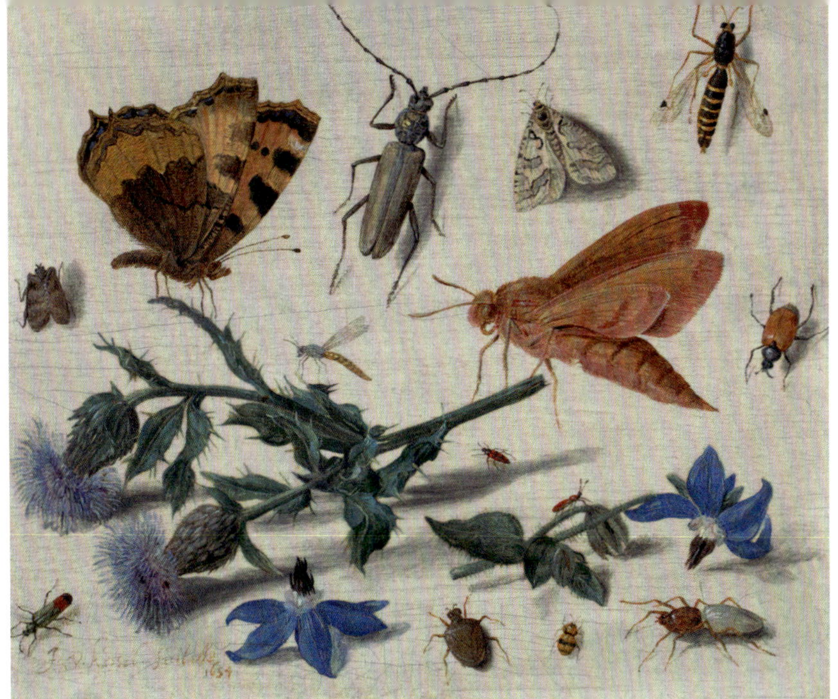

A GREEN HERB TART

Take some chervil, borage, parsley and some wild mint and rinse it in water. Press it well between two plates to get rid of the moisture. Finely chop or grate it. Take 125 g of fresh butter and melt it with the green herbs. Take six egg yolks and three egg whites and beat them well. Combine the braised herbs with the beaten eggs. Add some grated cheese, 125 g of sugar and some cream or milk, some cloves and lots of cinnamon.

***Brabants Kookboek*, second half of the 17th century**

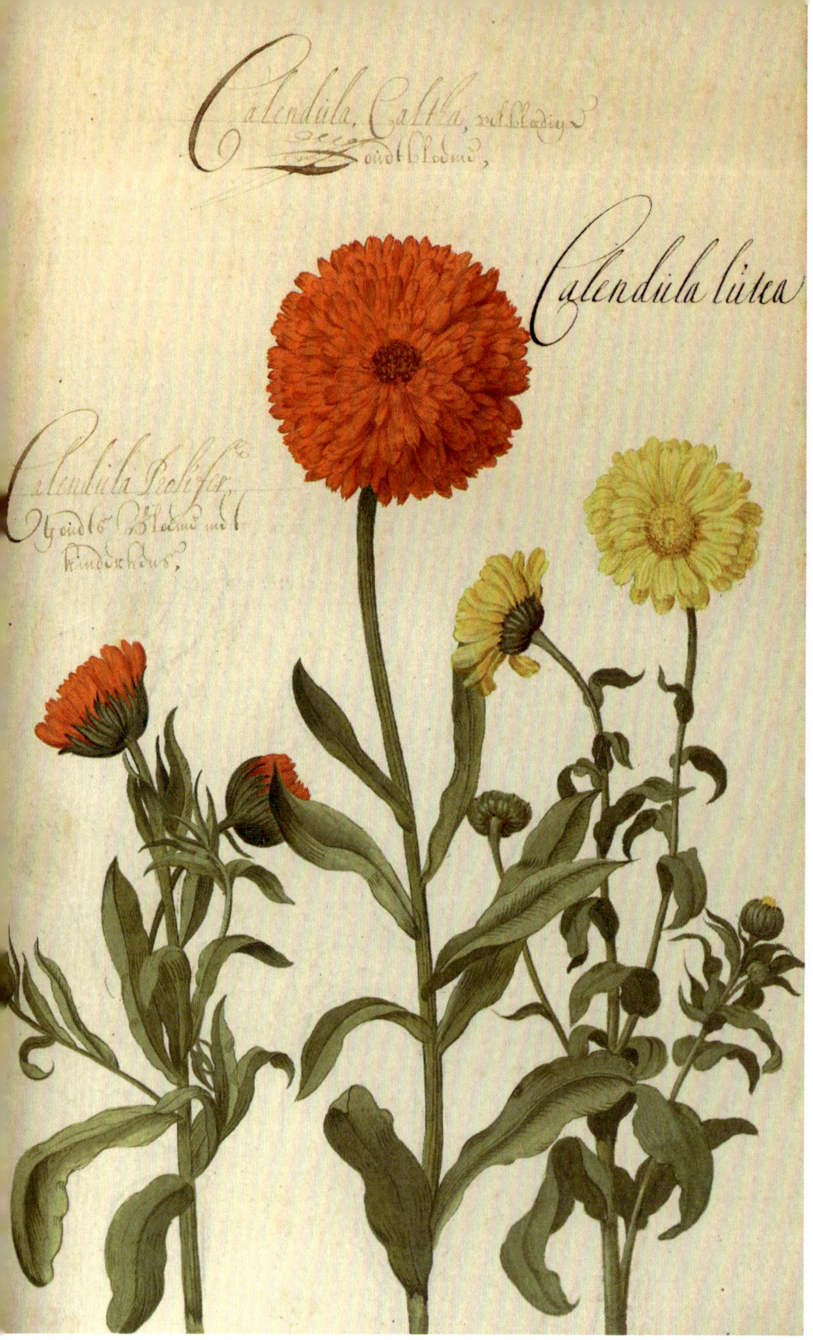

Franciscus de Geest, *Calendula* in the *Hortus Amoenissimus*, 1668, Biblioteca Nazionale Centrale, Rome

Calendula officinalis
Marigold

Calendula or marigolds flower in the beds of the Rubens Garden all summer long, until the first night frost. Lobelius describes these annuals as very beautiful, shiny or fiery flowers with many subtle and very small notches. They grew in all gardens back then, according to his herbal from 1581. Marigolds were used for medicinal purposes, as dye, and as a cooking ingredient.

According to almost all sources since the early Middle Ages, marigolds symbolise dependence on heaven or God, as the flower turns with the sun during the day. Ancient names of this flower refer to this, such as *Sonnenkruyt* (Sun herb), *Sponsa del Sola* (Fiancée of the Sun) and *Solsequiem* (Sun-follower). The widely-read *Den kleynen herbarius ofte kruydt boecxken* from 1640 states that marigolds 'rise with the Sun only to close at night when the sun sets'. In Rubens's day, several Antwerp chambers of rhetoric were named after plants, including the *Olijftak* (Olive branch), the *Violieren* (Stock flower), and also the *Goudbloem* (Marigold). The *Goudbloem*'s motto was *'Groeynde in deuchden'* (Growing in virtue).

The reworked edition of Dodoens's *Cruydt-boeck* from 1644 mentions how 'these pretty gold-coloured flowers are prized in gardens because they grow readily, producing a lavish display, in particular when they are laden with leaves or double flowers'. In the *Hortus Amoenissimus* Franciscus de Geest drew a yellow marigold and a nearly full orange marigold. According to the herbal, the seeds of the marigold are 'so crooked and tortuous that they resemble a bird's claws'. By 1644, marigolds came in different colours and shapes, 'single with a yellow or reddish or dark yellow-gold-coloured flower or also laden with leaves with double flowers. Some with other small blooms growing around a central flower'. The latter were called marigolds 'with children'. De Geest also drew them.

Like columbines, marigolds were used in wreaths or posies. Like today, calendula had many medicinal uses, for example to relieve menstrual pain, but also to stimulate menstruation. The Antwerp version of the *Den verstandighen hovenier over de twaelf maenden van 't jaer* from 1672 mentions that eating lots of marigolds promoted good eyesight. Marigold juice 'kills the worm in the ears', while marigold vinegar is said to be good in times of plague. Powdered dried leaves could soothe a toothache when applied to a cavity. Dodoens also mentions two other uses: the flowers and leaves of calendula imparted a nice taste and smell to meat sauce, and the plant was also used as a hair dye. The flowers dyed the hair yellow, while the leaves were used to produce a very green hair dye.

RECIPE FOR MARIGOLD PRESERVE

Combine 125 g of marigolds and 475 g of sugar, grind with a mortar and pestle, and put it in the sun.

***Brabants Kookboek*, second half of the 17th century**

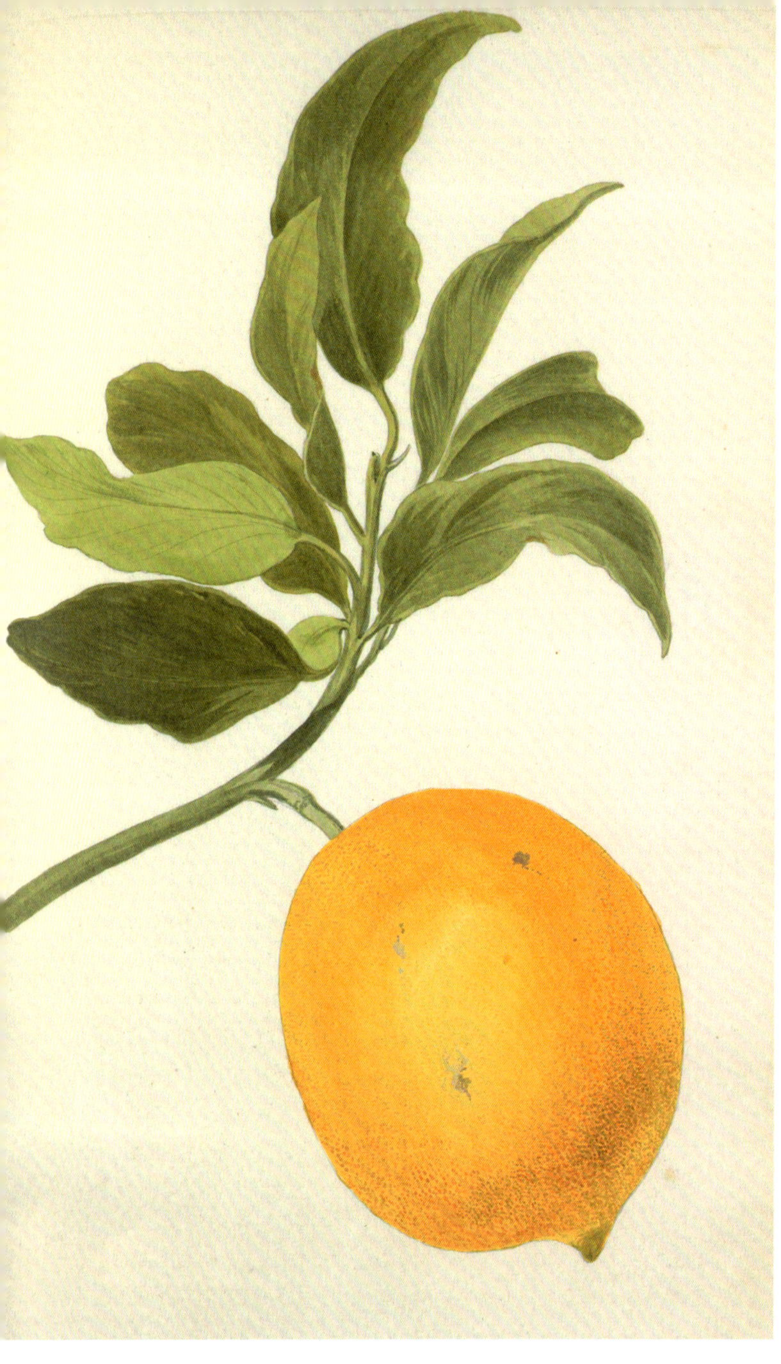

Pieter Holsteyn II, *Citrus medica* in *Flores a Petro Holstein ad vivum depicti*, 1640s, RHS Lindley Collections, London

Citrus

Many citrus plants such as lime, lemon, and orange look resplendent in the large oak tubs and colourfully glazed flower pots of the Rubens Garden. Thanks to a painting and archival documents, we know that Rubens owned an extensive collection of citrus plants and that he hired a cooper to make tubs in which the plants were placed. The same has been done in the 'new' garden. The current tubs are a superb example of the exceptional craftsmanship of Marleen Bonami, the only artisanal cooper in Belgium. The prickly *Poncirus trifoliata* (trifoliate orange) is the only citrus plant to survive winters in our climate, which is why it has been given a place in the parterres. Owning citrus plants was not uncommon or exceptional in Rubens's day, even in Antwerp. Citrus fruit was used for cooking and had various medicinal uses, including against the gout that Rubens suffered from. These are all the species you can find in tubs and pots in the garden from May until November: *Citrus aurantifolia* (lime), *Citrus aurantium 'Salicifolia'* (bitter orange), *Citrus aurantium 'Striata'* (a lemon and sour orange hybrid), *Citrus japonica* (kumquat), *Citrus latifolia* (Persian lime), *Citrus limon* (lemon) and *Citrus limonia Osbeck* (blood lemon), *Citrus mitis* (calamondin orange tree), *Citrus paradisi* (grapefruit), *Citrus sinensis* (sweet orange tree) and *Citrus volkmeriana* (red lemon).

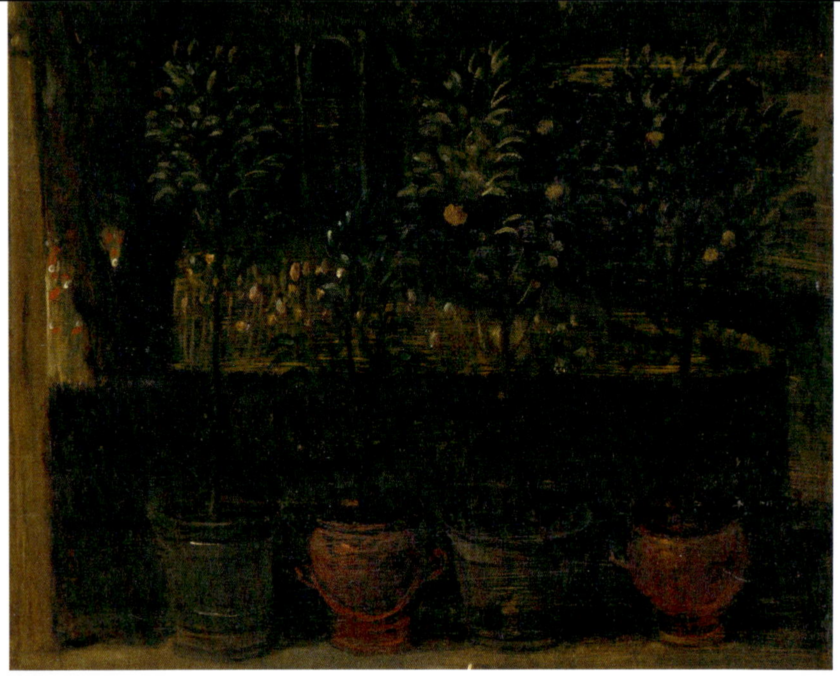

We know for certain that Rubens had a sizeable collection of citrus plants. The painting *The Walk in the Garden* depicts the artist, his wife Helena Fourment, and Nicolaas, Rubens's son from his first marriage to Isabella Brant, as well as an old woman feeding a pair of peacocks. The tulip meadow in the background is in full bloom, with a gurgling fountain. This was where Willem the gardener reigned supreme. Rubens accompanies his young wife to a cooler with a pouring jug and a set table in the garden pavilion. Rubens's own garden clearly provided the inspiration for the painting, which Rubens produced with the collaboration of his studio at the time of, or shortly after, his marriage to Helena Fourment on 6 December 1630. Four small trees in wooden tubs and glazed pots with ears line the path. The rightmost specimens bear orange fruits and can be identified as orange trees. This was the work of Jaspar the gardener.

DETAIL FROM Peter Paul Rubens (atelier), *Peter Paul Rubens, Helena Fourment and Their Son Nicolaas Walking in Their Garden ('The Walk in the Garden')*, c. 1630–1631, Bayerische Staatsgemäldesammlungen – Alte Pinakothek, Munich

Besides *The Walk in the Garden*, two archival documents are evidence that at least 13 orange trees and a lime tree grew in Rubens and Helena's garden. A recently discovered *'Memorie voor Vrou helena froment we(duwe) van wylen Heer P. Paulo Rubbens riddere over tgene by haer inden selve sterfhuyse ingecocht heeft inde maenden van april ende mey 1642'* (Memorandum for Lady Helena Fourment, widow of the late Sir Peter Paul Rubens, knight, on what she has acquired from the estate of the deceased in the months of April and May 1642) lists paintings and sculptures as well as four orange trees and a lime tree and the price that was paid for them. The prices for three large orange trees ranged between eight and 15 guilders; a smaller *'boomke'* cost two guilders and 12 stuivers. These prices are considerably higher than the price people would have paid for fruit trees, which cost less than a guilder. On 27 March 1659, Moretus paid his gardener seven guilders and eight stuivers for eight apricot trees, one guilder for two plum trees, and 15 stuivers for one cherry tree.

Perhaps the oranges and the lime remained in the garden of the Rubenshuis for some time after 1642. In 1645, Rubens's young widow Helena Fourment married Jan Baptist van Brouchoven. The couple lived in Antwerp for some time. However, from 1648, Helena rented the house and garden to William and Margaret Cavendish. Shortly after 1650, Helena and Jan Baptist and their family moved to Brussels – possibly surrounded by the citrus plants from Helena's beloved garden in Antwerp.

Besides the 'memorial' for Helena, the *Staetmasse* or settlement of Rubens's estate in 1645 also mentions citrus plants in the garden. This archival document cites a payment of twenty guilders to Jaspar the gardener for a year's pay for minding the orange trees in the garden. A further three guilders were also

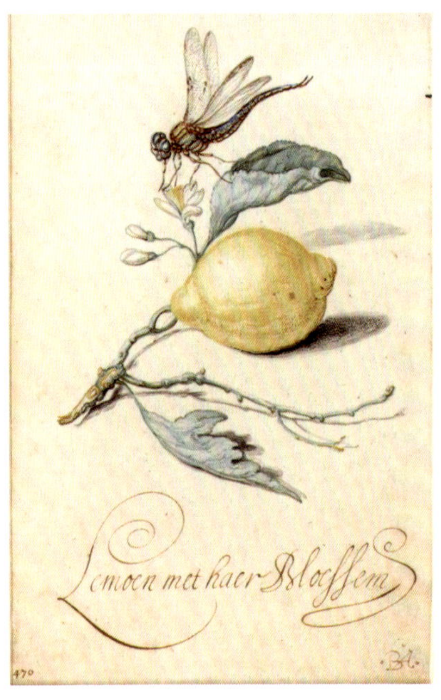

Baltasar van der Ast, *Lemoen met haer Bloessem*, 1632–1657, private collection

paid for three new tubs for the orange trees. Elsewhere in the deed, another payment of 15 guilders is mentioned 'to the cooper for ten tubs for orange trees'. Rubens and his gardeners thus hired a cooper to plant the orange trees in the appropriate manner.

The name of Rubens's cooper is not mentioned in the *Staetmasse*. Thanks to paintings and prints of that period, we have a good idea of what the tubs would have looked like: they were straight wooden tubs with simple iron mounts (at the top and bottom, and sometimes in the centre). A good example of an orange tree in a wooden tub can be seen in the aforementioned work from 1651 by David Teniers II, depicting an elegant company in a garden with the Rubens-inspired pavilion. They are also depicted on the title page of Franciscus van Sterbeeck's *Citricultura oft regeringhe der uythemische Boomen te weten Oranien, Citroenen, Limoenen, Granaten, Laurieren en andere,* the foremost handbook on citrus cultivation in Flanders – published in Antwerp in 1682. The tubs were occasionally fitted with handles to facilitate transport during the winter months.

Orange, lime, and pomegranate trees were common in Antwerp city gardens in Rubens's day, as archival research has shown. Nicolaes Rockox – a friend and patron of Rubens – had, among other things, ten orange trees, with some bearing oranges, sitting in the cellar of his house on Keizerstraat in December 1640. Some young orange trees also overwintered in Rockox's cellar. In the spring of 1641, a notary recorded, among other things, that there were two pomegranate trees and four orange trees in the garden of Alderman Charles de Tassis behind the Capuchin convent. The 1645 inventory of Gaspar Charles, a patron of Rubens, lists nine orange trees with tubs, some beds with orange trees, and two pomegranate trees in his garden on Lange Nieuwstraat. In February 1652, there were several trees in the flower garden of the Antwerp alderman, art collector, and bookseller Jan van Meurs on the Hopland near Rubens's house, including three pomegranate trees, a large orange tree, and six tubs with young orange trees. The French diplomat and physician Balthasar de Monconys wrote the following about Rubens's neighbours, the Duarte family who lived on the Meir, in his travel diary in July 1663: *'Nostre Marchand Dom Gilles mena M. le Duc chez un Portugais nommé Doüart, qui a un beau Iardin, & les orangers les mieux formez qu'on puisse voir'* (Our merchant Dom Gilles brought His Grace the Duke to a Portuguese man named Doüart, who has a lovely garden and some of the best orange trees in the city). In the garden of Rubens's granddaughter Helena-Françoise and her husband Jan-Baptist Lunden, the Antwerp citrus specialist Franciscus van Sterbeeck saw a grooved, flaming orange-apple tree hybrid, with beautiful mature fruit.

But other Antwerpians – whom Rubens may not have known – also had extensive collections of orange trees. On Korte Nieuwstraat, as many as thirty-four small and large orange trees adorned the garden of Maria van Houte and Jan Liebrechts's house *De Gouden Huif* in 1652. Merchant Olivier van Houte, who lived on Everdijstraat, had a display of 26 orange trees and nine pomegranate trees in pots in February 1654. At the home of Pascqual de Decker, who was the amman or bailiff of Antwerp,

NEW TUBS AND FLOWER POTS

The tubs in the current Rubens Garden were made by Marleen Bonami. In her workshop in Nevele, she crafts straight oak tubs in the old-fashioned way. A smith supplied the mounts and handles. This niche specialisation is no longer offered by woodworkers, joiners or the former 'coopers' in Belgium.

The green and brown-glazed flower pots come from Anduze, a village in southern France known since the 17th century for its orangery pots. At the poterie '*Le Chêne Vert*' – which obtained the 'Entreprise du Patrimoine Vivant' label – antique pots are restored, and new pots are turned and glazed according to Baroque models using traditional methods. This is done by a small team of specialists.

For all kinds of trees, that are not frost-resistant and have to be stored inside the house, wooden tubs are better than stone pots, for the simple reason that the stone is too cold in winter, which causes the tender roots, which sprouted in the last days of summer, to suffer great inconvenience and disadvantage.

Franciscus van Sterbeeck, *Citricultura*, 1682

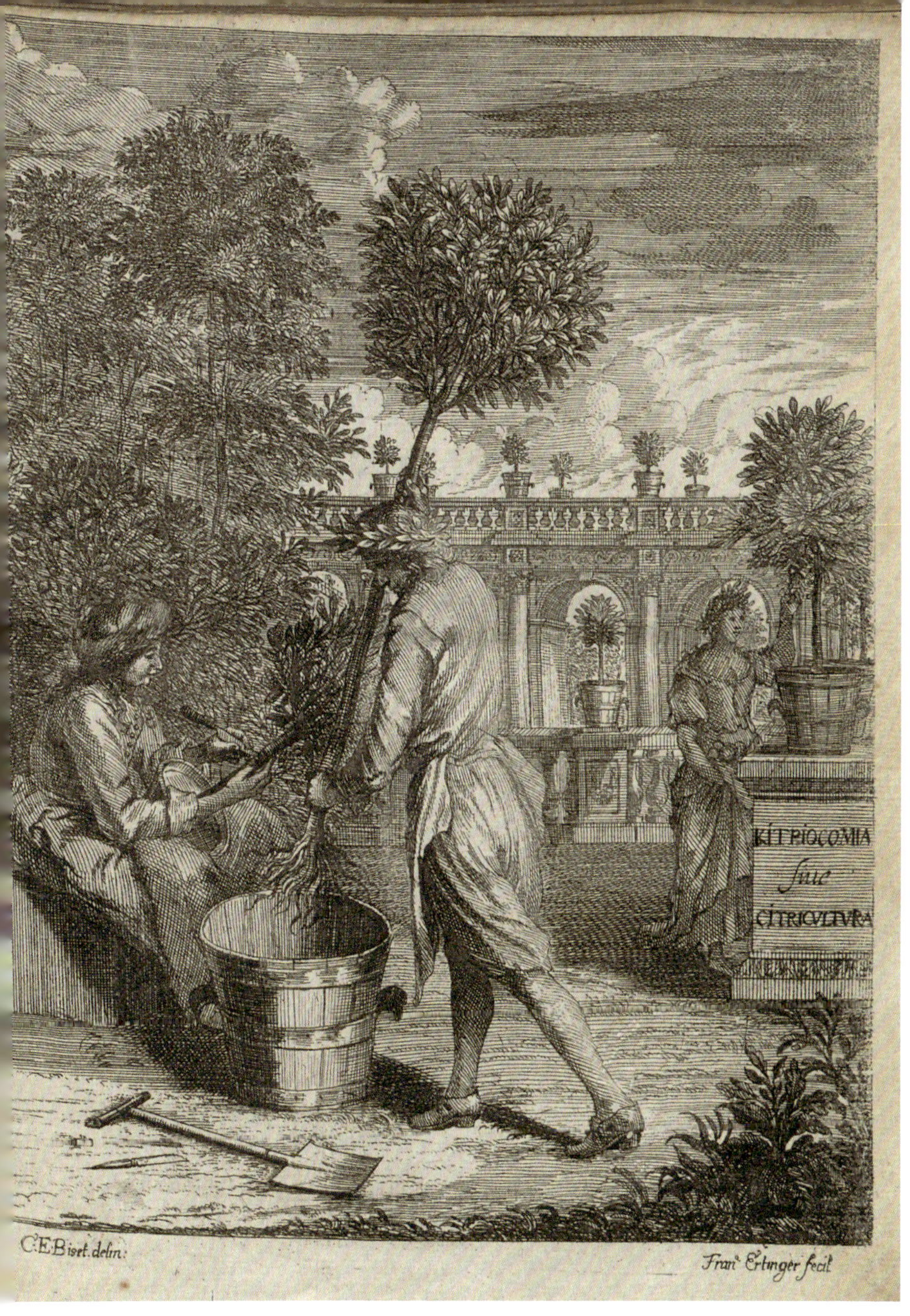

Charles Emmanuel Bizet and Franciscus Ertinger, Title page for the *Citricultura* of Franciscus van Sterbeeck, 1682, Hendrik Conscience Heritage Library, Antwerp

Franciscus van Sterbeeck saw a wild or naturalised orange tree which he described as ‘a clean straight-stemmed tall wild orange tree, with a trunk that was as thick as a child’s forearm, which his mother had planted or sown from the seeds of an orange and which was still fully wild. Without being grafted or oculated, as attested by its many sharp thorns and other signs’. According to Van Sterbeeck, the tree was forty years old and was bearing fruit for the fourth year. ‘This is the first and only wild orange tree in all of the Netherlands, the merit of which goes entirely to our Antwerp air’, he concluded.

However, most citrus plants in Antwerp were joined to rootstock and were thus propagated by grafting (an entire branch is attached) or oculating (only the ‘eyes’ are joined to wild rootstock). According to Van Sterbeeck, orange, lime, and white jasmine trees were not shipped from Italy to Antwerp until 1668. They came from San Remo by way of Genoa, where ‘the most experienced gardeners and practitioners in all of Italy’ could be found. The climate there was similar to ours, which is why trees from this region did better than trees from Spain, Portugal or hotter places, Van Sterbeeck said. Orange trees were also shipped from Antwerp to Friesland, Denmark, Sweden, and other places. According to Van Sterbeeck, Italian wooden crates with holes were best suited for transportation.

The same Franciscus van Sterbeeck described how the limes in his garden were the same fruit offered for sale in Antwerp from Spain. But there the fruits were picked while still unripe and green; they ripened during transport, which he thought was unhealthy. Every year, the Antwerp trees produced many very ripe fruits that were *aenghenaemer* and *ghesonder* (more pleasant to eat and healthier). While citrus fruit appear in all kinds of local recipes, the plant also had many medicinal uses that may

RIGHT Citri in a coloured copy of Besler’s *Florilegium*, Teylers Museum, Haarlem

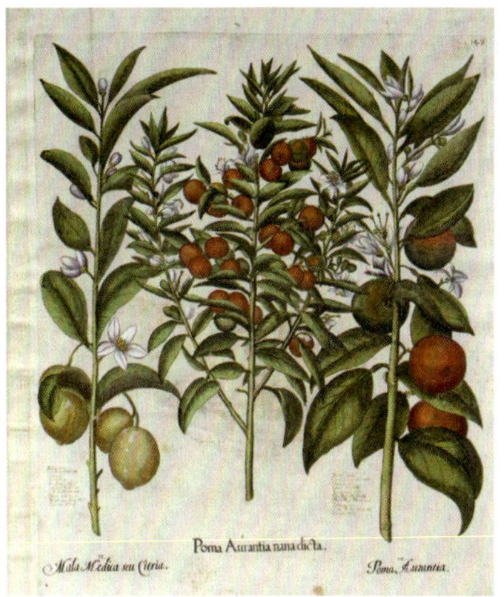

have been known in Rubens's household. Orange blossom oil was used for gout pain, for example, the painful rheumatic inflammations in hands and feet from which Rubens suffered. A mask with lime juice reduced spots, freckles, and wrinkles on the hands and face, and removed ink stains from book covers.

HOW TO MAKE GOOD LIME WINE

Take half a pound of lime peel, 1/2 ounce of cinnamon, 1/4 liquorice, 1/4 raisins, 1/4 figs, 65 g of anise, six jugs of wine and half a gram of saffron.

***Antwerps kookboek van Clara van Molle*, first half of the 17th century**

Cyclamen hederifolium

Sowbread

This plant, which flowers low to the ground, is a real show-stopper from late summer until the first frosts, with its foliage with a serrated edge, silver-spotted and speckled leaf pattern, and the elegant petal structure of the exquisite pink flower. After flowering, the most wonderful foliage appears that stays green all winter long. Cyclamen were a coveted species in early 17th-century flower still lifes, where they often appear at the bottom of the bouquet. The flower had several medicinal uses. The tuber was deemed dangerous for pregnant women.

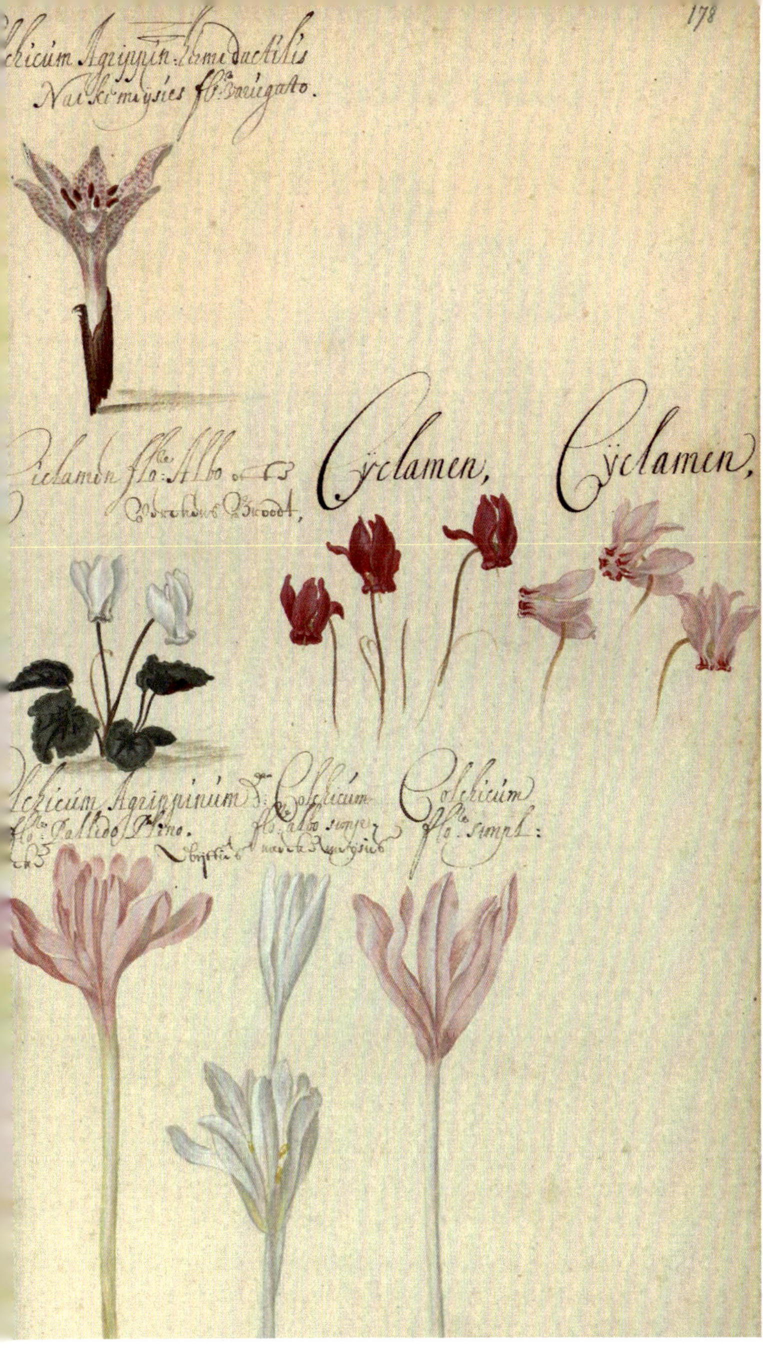

Franciscus de Geest, *Cyclamen hederifolium* in the *Hortus Amoenissimus*, 1668, Bibliotheca Nazionale Centrale, Rome

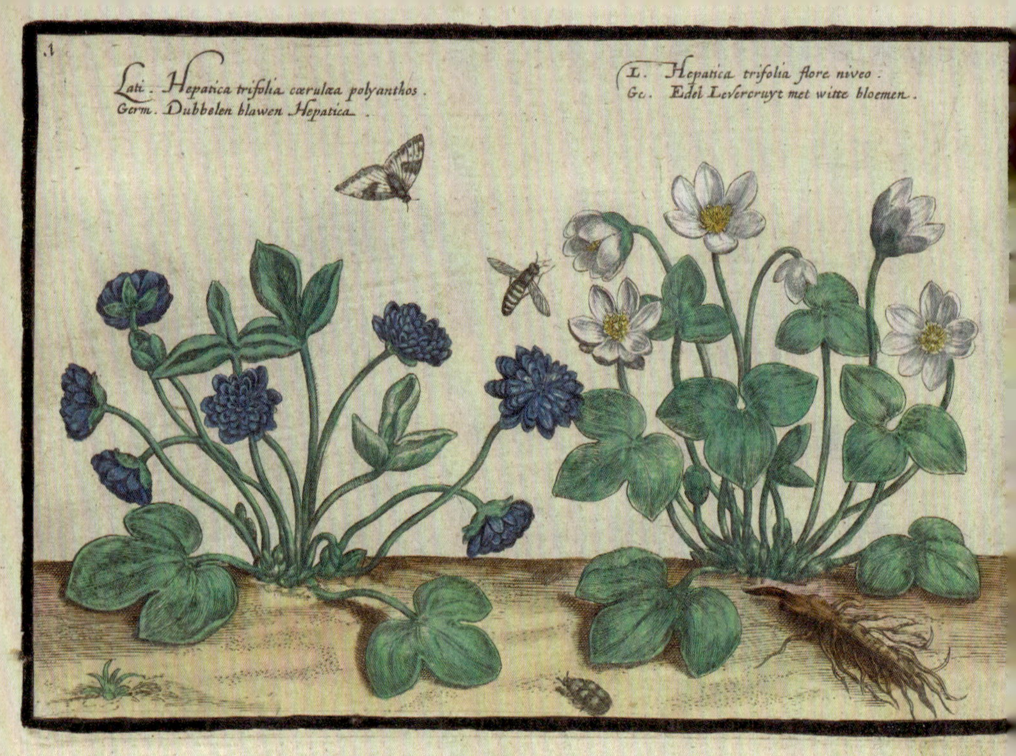

According to Dodoens's *Cruydt-boeck* of 1644, the flowers 'bloom on very slender stems, arching down, although the petals point up, with the purplish colour of blue violets, but not very dark, shiny, with almost no scent'. According to the *Cruydt-boeck* , this species of cyclamen grew in the Netherlands and on farmland in the border region of France, Flanders, and Artois. Sowbread was a wreath herb, used to decorate posies.

Sowbread had many uses in Rubens's day, from curing haemorrhoids and cataracts to improving bowel movements and

Crispijn de Passe II, *Hepatica* from the *Hortus Floridus*, 1614, Oak Spring Garden Foundation, Upperville, Virginia (USA)

preventing hair loss. Pharmacists called the flower with the imposing, round and flat root tuber *Panis porcinus*, while it was colloquially known as *verckens-broot* or *seugen-broot* (sowbread). The tuber of sowbread is also depicted in botanical handbooks and early florilegia, including Besler's *Hortus Eystettensis* (see elsewhere in this book on pages 58–59). It could be used as an ingredient in biscuits to render love potions harmless, and in ointment to heal the scars of smallpox and measles. The tuber was deemed dangerous for pregnant women. They were not allowed to ingest it or even touch it. According to Dodoens, even stepping on the root was enough to cause severe problems. The English herbalist John Gerard reportedly even fenced off the plants in his garden to avoid premature births. In the garden of Charles de Tassis in Antwerp, a bed with *'dobbelen apatica ende folio hedre'* had been laid out in 1641. So the unassuming cyclamen were paired with *Hepatica* or refined double liverworts. In the Brussels garden of a painter, *'Folhederen witt de blomme'*, *'Folhederen met Incarnaete blomkens'* and *'Pannis porzinons'* bloomed in 1609 (ivy-leaved cyclamen with red and white flowers and sowbread). Franciscus de Geest drew sowbread in different colours in the *Hortus Amoenissimus*.

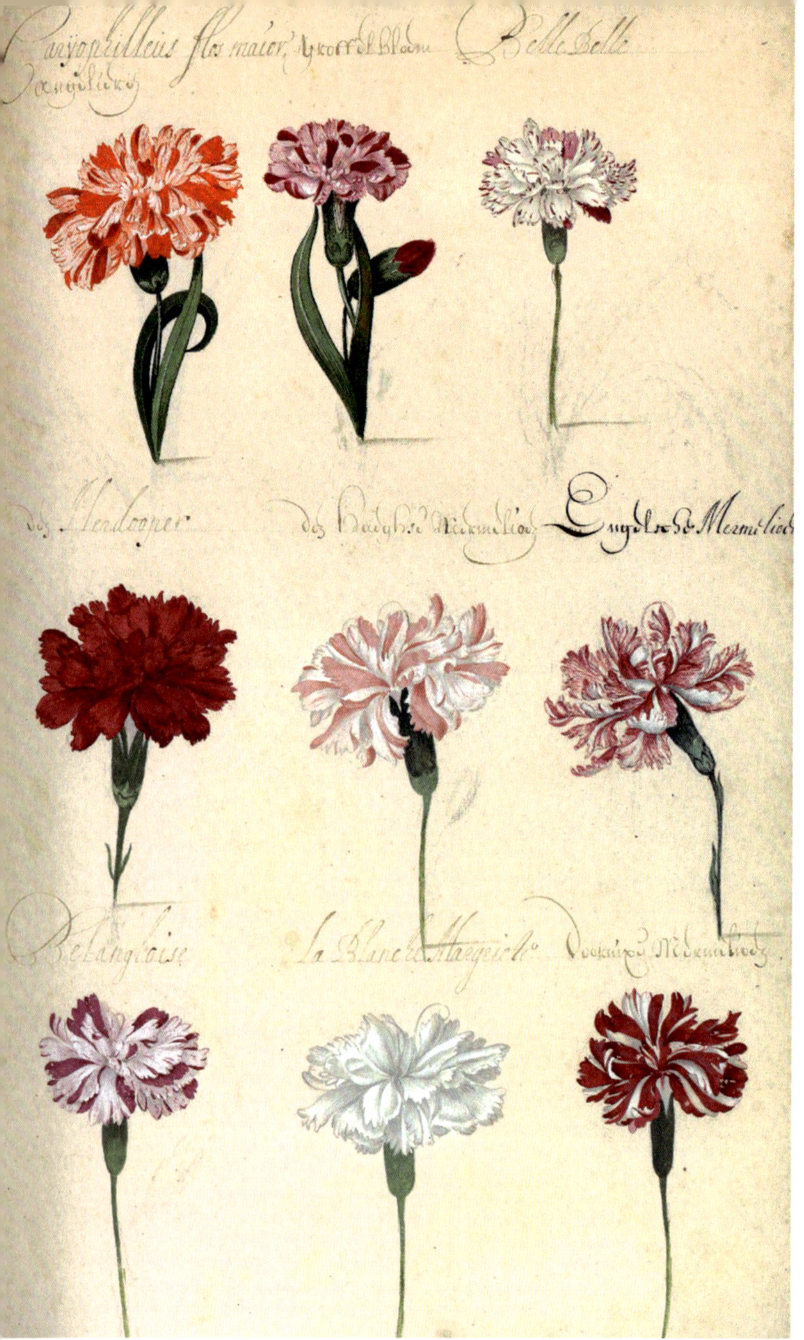

Franciscus de Geest, *Dianthus* in the *Hortus Amoenissimus*, 1668, Biblioteca Nazionale Centrale, Rome

Dianthus

Carnation

Carnations or pinks were among the most popular garden flowers in Rubens's time, because of their beauty and scent. Both the Carthusian pink *(Dianthus carthusianorum)* and the maiden pink *(Dianthus deltoides)* were given a place in the beds of the new Rubens Garden. The 1672 edition of *Den verstandighen hovenier over de twaelf maenden van 't jaer*, published in Antwerp, describes how the juice of the flowers has analgesic properties when used for head wounds, and how preserves of carnation flowers strengthen the heart. Carnations were kept indoors in pots or tubs in winter to protect them against the cold. This explains why they are often listed in estate inventories, which traditionally focused on household effects rather than plants in gardens.

Jenoffelen or large carnations are described in Dodoens's reworked *Cruydt-boeck* of 1644 as 'clean and lovely to look at, emerging from long, round serrated or jagged pods'. There were double and single forms. Sometimes pale purple, dark purple-red, brown, but also white, white and purply, or pink, incarnate and sometimes speckled with various colours. A colourful collection of carnations can be found in Franciscus de Geest's *Hortus Amoenissimus*. *Pluymkens* (little feathers) were smaller than large carnations. Their flowers, which were more serrated on the sides or more ruffled were very similar to birds' feathers in appearance, hence their name. Large carnations and *pluymkens* are depicted in Adriaen Collaert's *Florilegium*, a model book published by Philips Galle in Antwerp as early as 1590.

In 1676, H.I.B. Reyntkens, a monk of St Peter's Abbey in Ghent, published *Den sorgvuldighen hovenier*, a garden manual that was very widely distributed. In it, Reyntkens describes the carnations as *'Angelieren, oft Jenoffels diemen Potbloemen noemt'* (carnations which are called pot flowers). They were popular flowers, 'not just because of their beauty but also for their lovely scent'. Dodoens describes the smell as sweet, Lobelius compares the smell of some full-grown specimens to cloves. Carnations were widely used in flower arrangements and wreaths. When pickled in vinegar, the flowers had a very lovely taste and pleasant smell. Because of their scent, carnations are a common flower in allegorical representations of the senses, as in the *Allegory of Smell* painted by Jan Brueghel I and Rubens around 1617.

Carnations were kept in pots or tubs to protect them from the winter cold. They are often depicted as such in prints and paintings, including in the foreground of the painting *The Walk in the Garden*. Often the flowers are supported by a wooden trellis. Dodoens's book of herbs recommends storing large carnations

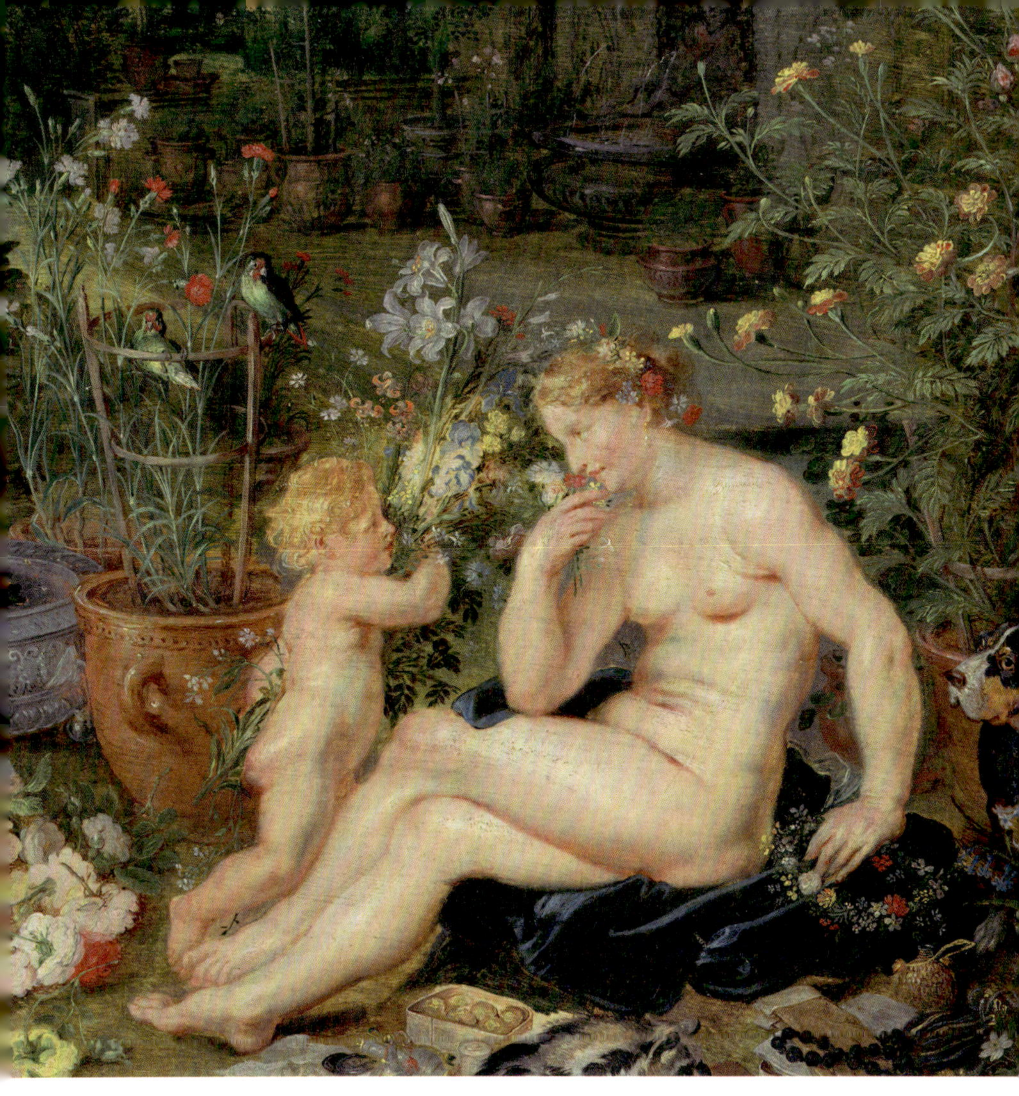

DETAIL FROM Jan Brueghel I and Peter Paul Rubens, *Allegory of Smell*, c. 1617, Museo nacional del Prado, Madrid

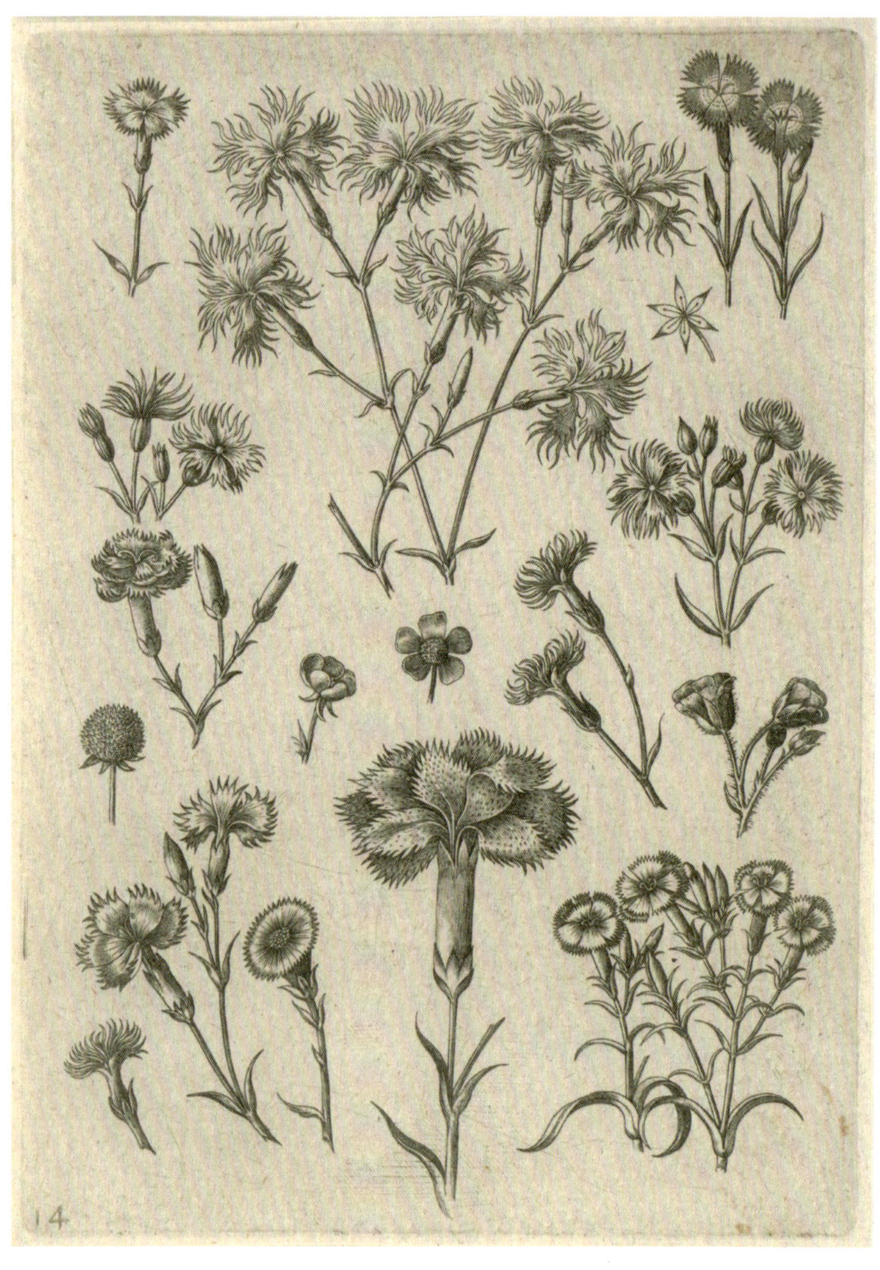

Adriaen Collaert, *Dianthus* in the *Florilegium*, c. 1590, Rijksmuseum, Amsterdam

in tubs or containers in the wine cellar or in lukewarm places during winter. Even there they will continue to flower, regardless of how harsh the winter is. They are often listed in estate inventories, precisely because the plants overwintered indoors and were sometimes even stored there permanently. In the front attic on the street side, Antonette Wiael, the widow of art dealer and panel maker Jan van Haecht, had stored 21 tubs with carnations in 1627. The 1646 settlement of the estate of Cornelis de Wael, a merchant in glass, mirrors, and ebony, records how the tubs with carnations from the garden of the house *Veneetsen Spiegel* in the Steenhouwersvest, had been sold at public auction in the Beurs and on the Vrijdagmarkt, together with orange trees, other trees, and furniture, for a total of 153 guilders and 18 stuivers.

Not all carnations were kept in pots or tubs, however. In the spring of 1641, an Antwerp notary recorded 'a bed with large carnations' in the garden of Alderman Charles de Tassis behind the Capuchin monastery. In 1637, Canon Antonio de Tassis of the Cathedral of Our Lady had sold 'a pound of carnations for six guilders per pound' to a group of Antwerp tulip collectors that included Rubens's brother-in-law Peter Hannekart. He had gifted a quarter pound of carnations to Cornelis van Engelen and Emanuel Verspreet.

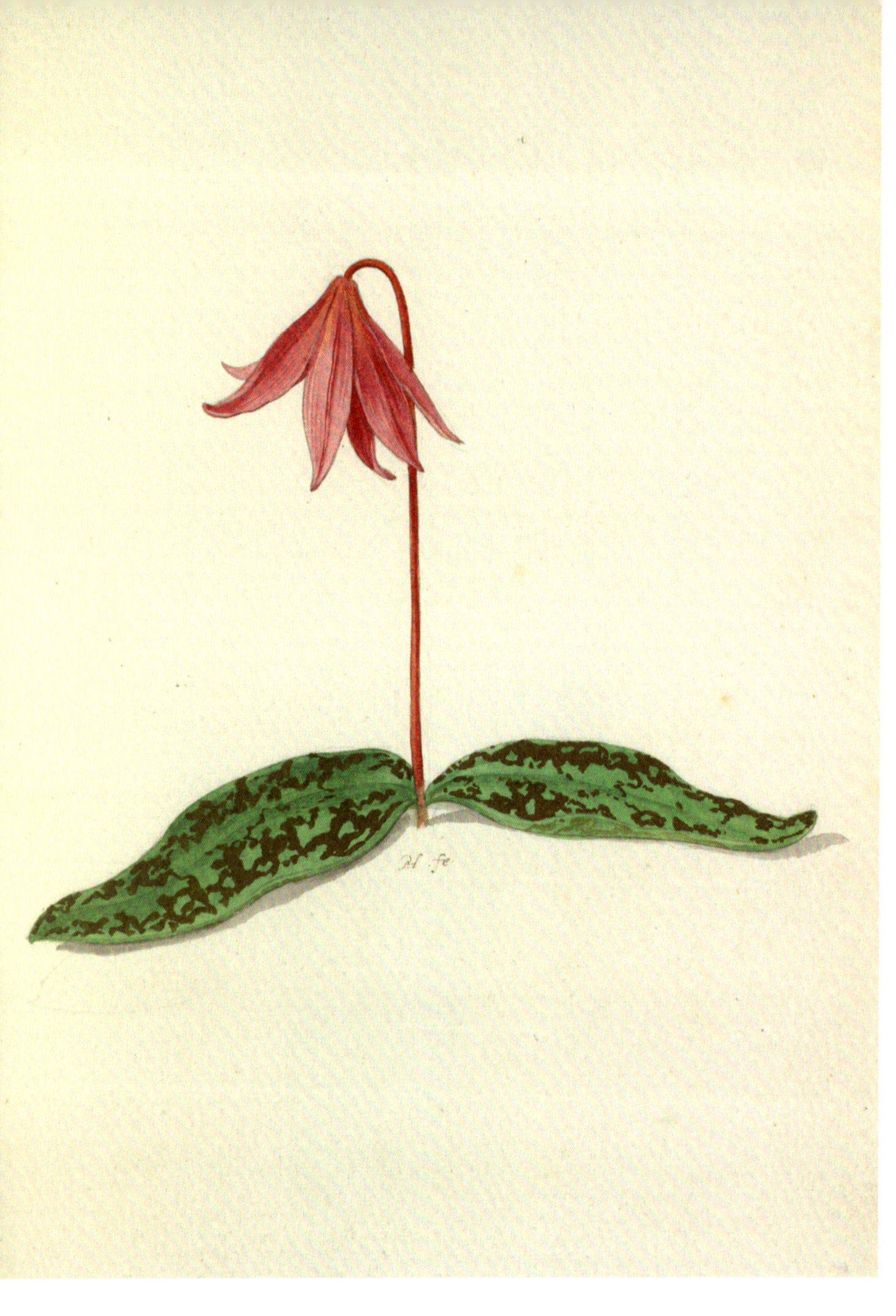

Pieter Holsteyn II, *Erythronium dens-canis* in *Flores a Petro Holstein ad vivum depicti*, 1640s, RHS Lindley Collections, London

Erythronium dens-canis

Dog's tooth violet

In the beds of the new Rubens Garden, pink *Erythronium dens-canis* and golden yellow *Erythronium dens-canis 'Pagoda'* bloom in April. The plant owes its name (dog's tooth violet) to its narrow, elongated bulb, which resembles a dog's canine tooth. In Rubens's day, this was an exclusive flower whose leaves and root had various medicinal uses.

The unusual *Erythronium dens-canis* is already meticulously described in Dodoens's *Cruydt-boeck*: 'In early spring, dog's tooth violet with purple flowers usually produces two, rarely three leaves'. When the mottled or marbled leaves emerge, a flower appears 'standing on a plain, hollow and slender stalk, with six elongated petals that curl downwards, opening when the sun's heat can be felt and its leaves pointing straight up', just like with sowbread. Dodoens described the flower itself as 'very pretty, pale purple in colour, with six purple threads or stamens within, and decorated with a white three-sided sepal'. Dog's tooth violets belong to the stinzen plant family, a collective name for a special group of naturalised spring flowers. They are mainly bulbous and tuberous plants, and root crops that were planted from the 16th century on country estates, around castles, and manor houses.

This exclusive plant pops up here and there. In 1609, Charles of Arenberg noted how red and white dog's tooth violets flowered in the garden of a painter friend in Brussels. Pieter Holsteyn II immortalised the dog's tooth violet, with striking marbled leaves, in the 1640s in the flower album *Flores a Petro Holstein ad vivum depicti*. *Erythronium dens-canis* was also included in the 1633 herbal of pharmacist and Norbertine friar Bernard Wynhouts, the second oldest herbal from our region to survive. Native and exotic species were cultivated in the garden of Dielegem Abbey. In the album, dog's tooth violet was dried along with the leaves of *Datura* and *Daucus* (thorn apple and wild carrot, both of which are also found in the Rubens Garden). Dodoens describes how a powder or flour of dried roots in the porridge of young children can cure stomach worms and epilepsy. Taken with wine, the root cleansed the gut and also served as a fortifying tonic for the entire body.

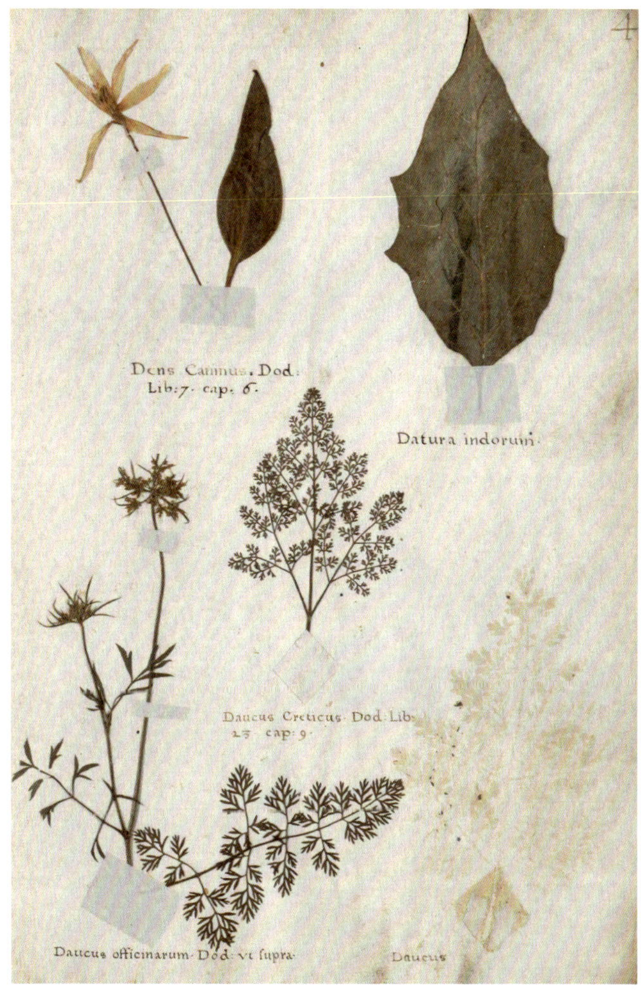

Bernard Wynhouts, *Erythronium dens-canis* in the *Herbarius continens species plantarum tum patriarum tum exoticarum ad vivum prout nascuntur in horto infirmariae celeberrimae Abbatiae Diligemensis*, 1633, University Library, Ghent

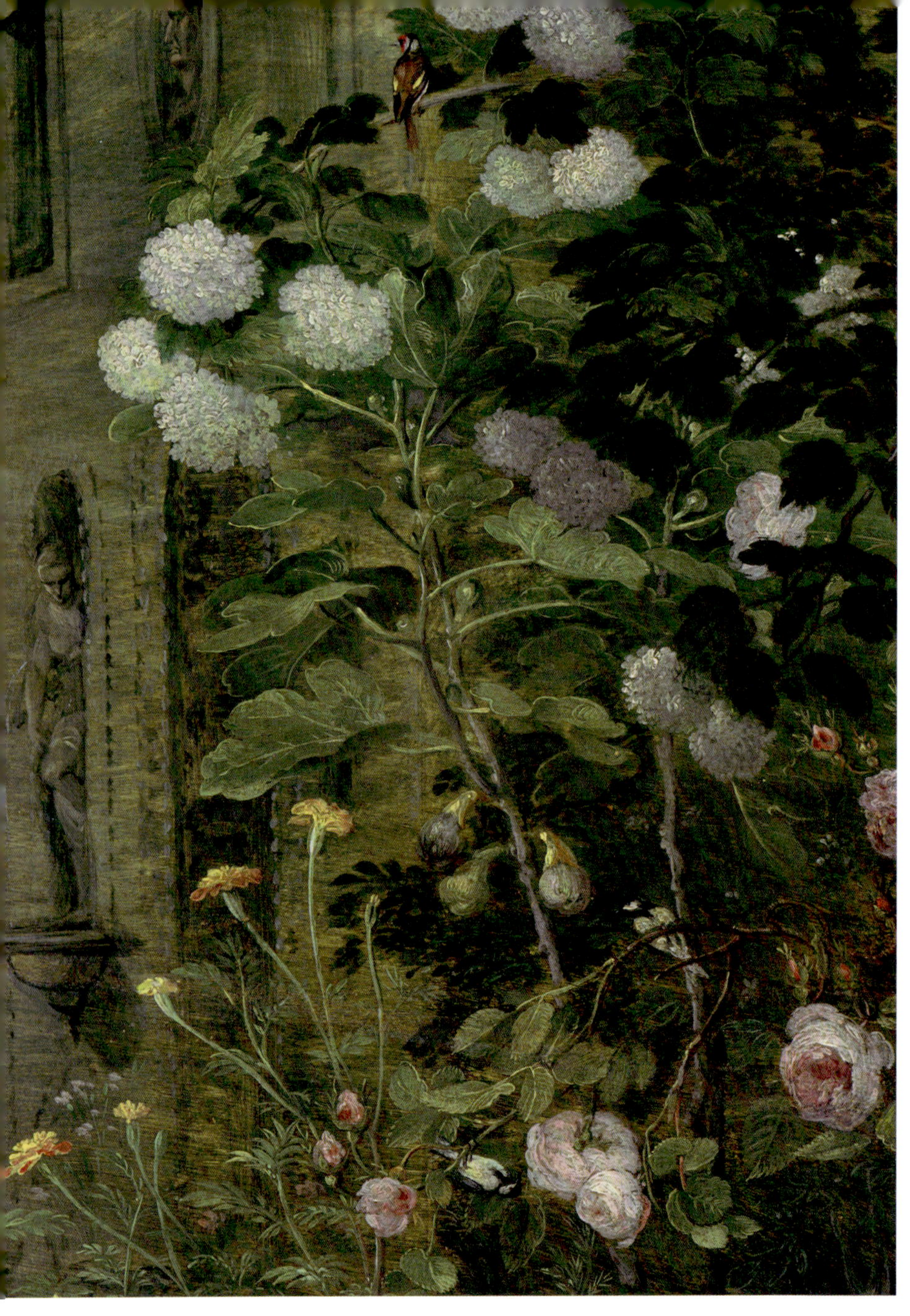

DETAIL FROM Jan Brueghel I and Peter Paul Rubens, *Allegory of Smell*, c. 1617, Museo nacional del Prado, Madrid

Ficus carica

Fig

A fig tree proudly spreads its leaves in the parterre in front of the pavilion. This is one of the few plants that we know with absolute certainty grew in Rubens's garden, thanks to written sources. Rubens bought at least 30 fig trees from pharmacist Peter van den Broeck in the 1630s. Southern fig trees were not that unusual in Antwerp city gardens in Rubens's day, although they needed lots of warmth and sun, and mild winters. To avoid freezing, they overwintered indoors in pots or tubs. Parts of fig trees and their fruits were used for many ailments, including against the gout from which Rubens suffered at times from as early as 1623.

In the summer of 1638, Rubens and his family stayed at *Het Steen*, his country house in Elewijt. Lucas Faydherbe, a 21-year-old sculptor from Mechelen and a close friend of Rubens, had been appointed as caretaker in Antwerp by the artist. Faydherbe had to ensure that the studio was operational and that nothing disappeared from it. In a letter from Rubens to Faydherbe dated 17 August of that year, the artist asked him to bring a panel featuring three tronies from Antwerp and some bottles of *vin d'Ay*, as the stock he had brought with him to Elewijt had run out. In a postscript to the letter, Rubens also asked that Willem the gardener be reminded of the *Rosilepeyrkens* (Rosile pears) and figs, should there be any, or any other delicacy from his garden. Thanks to this note – just a quick scribble on paper basically – we know that figs and pears were growing in Rubens's garden in August 1638. Moreover, this is Rubens's only written testimony about his garden.

Please check when you leave that everything is locked away and that there are no originals or sketches in the studio. Also remind Willem that he should send us, when he has the time, some Rosile pears *(Rosilepeyrkens)* and figs if there are some, or any other delicacy from the garden. Come as soon as you can, so the house can be closed because as long as you are there, you cannot lock out the others.

Letter from Rubens in Elewijt to Lucas Faydherbe in Antwerp, 17 August 1638

In the *Staetmasse*, the settlement of Rubens's estate from 1645, we learn even more about the fig trees from the garden. An account by pharmacist Peter van den Broeck shows that he not only supplied Rubens with medicine for gout, but also that he had sold the painter 30 fig trees between 1632 and 1640. They cost a total of 100 guilders. Pharmacist Van den Broeck had a shop in the heart of Antwerp on the west side of Beddenstraat, adjoining the corner house on the Schoenmarkt. He was a family friend. His name is already mentioned in the settlement of the estate of Rubens's first wife Isabella Brant from 1628 in relation to medicines he supplied. He possibly grew fig trees on a patch of land he leased in the south of the city, on the east side of Kloosterstraat. Precious tulips were later stolen from his garden in 1646.

Dodoens's *Cruydt-boeck* of 1644 shows how the fig tree flourished widely in warm countries. Those who wanted to enjoy the fruit in our regions had to have a lot of patience, however. The plant could only reach 'perfection or maturity' in a warm sunny place, protected from northerly winds, and with mild winters or very warm summers.

Archival research shows that fig trees were not so unusual in Antwerp city gardens in Rubens's day. The cellar of gardener Jan van Beneden contained six fig trees in January 1620. At the home of tailor François Ghyger and Antonette Verbeeck on 1 February 1634, there was a tub with a small fig tree in the beer cellar of the *De Lelieboom* house in the Sint-Katelijnevest. Like citrus plants, Antwerp fig trees thus overwintered indoors and were therefore kept in pots or tubs. Citrus specialist Franciscus van Sterbeeck reminded readers of the changing seasons in his *Citricultura*, citing the year 1680 as an example. By then, he had picked his second harvest of ripe figs in Antwerp and explained how his fig trees still stood in the garden on 4 November, with other plants, and that they could remain in the garden without damage until the 25th of that month. It had only started to freeze on 1 December, Van Sterbeeck writes. In the pleasure garden of Antwerp alderman, art collector, and bookseller Jan van Meurs on the Hopland, near Rubens's house, several trees had already returned to the flower garden on 27 February 1652, including three fig trees. In the spring of 1641, a notary recorded 'a bunch of fig trees' in the garden of Alderman Charles de Tassis.

Reyntkens's gardener's handbook of 1676 shows how fig trees were sometimes planted in full soil under the vines. According to the same source, fig trees could also be used for hedges or planted against a wall to cover it. In case of cold and prolonged frost, 'it must be wrapped or it will die'. However, not all fig trees survived our cold winters. Among the many tubs with the aforementioned citrus plants, in the winter of 1652, a tub with iron mounts was found with a dead fig tree in it in the garden of the surgeon Benedictus van den Walle, who lived in a house called the *Swanenborch* on the Kipdorp.

The potency and effects of unripe, fresh, and dry figs and of the juice from young branches of the wild fig tree were praised in Rubens's day. The plant was recommended for colds, cystitis, constipation, ulcers, growths, itching, protruding ears, bee, scorpion, and snake bites, rabid dog bites, smallpox, measles, freckles and warts, among others. When combined with wax (or an egg yolk), fenugreek flour, and vinegar, the juice, fruits, and leaves could relieve gout pain, especially swelling in the joints of the foot. The mixture had to be added to a paste or bandage and was applied to the sore like a bandage. Whether Rubens used his more than 30 fig trees for this purpose is not known, but we know that he regularly suffered severe gout attacks from 1623 onwards.

RIGHT Jan van der Groen, *Lemon, fig*, and *orange tree* from *Den Nederlandsen hovenier*, 1669

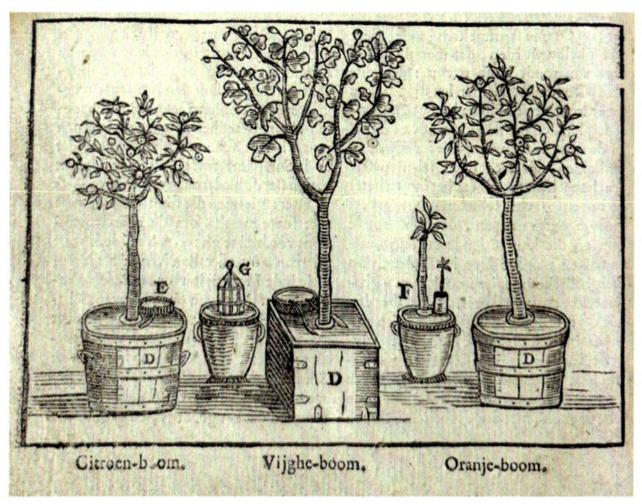

La bonté de la figue n'est mise en dispute, chacun tenant ce fruit-là estre des plus exquis, lequel & le raisin, par jugement universel, sont estimés la coronne de tous autres.

The goodness of the fig is not disputed; everyone considers this fruit one of the most exquisite, along with the grape, which everyone believes is better than all others.

Olivier de Serres, *Le théâtre d'agriculture et mesnage des champs*, 1600

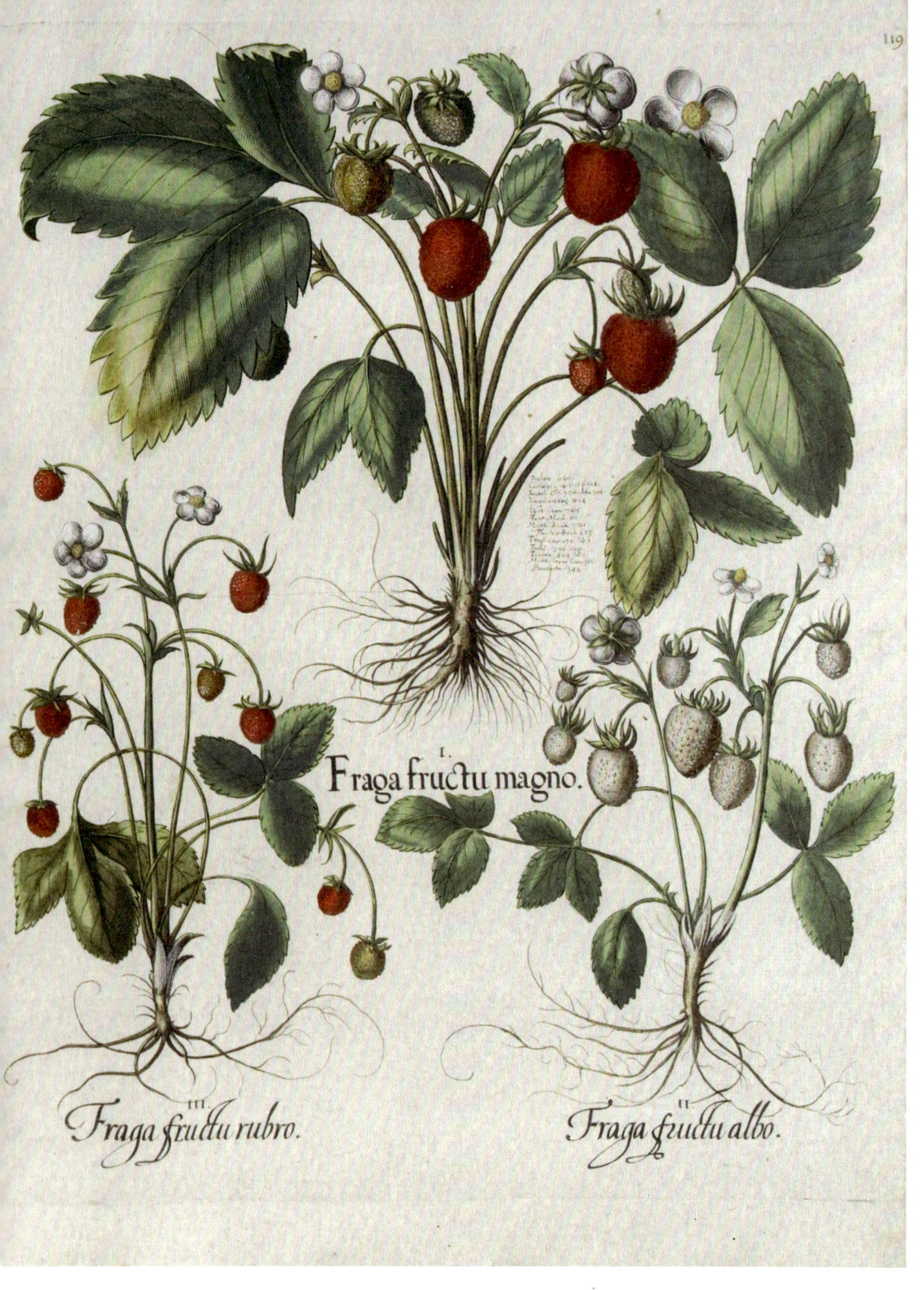

Fragaria vesca in a coloured edition of Besler's *Florilegium*, 1613, Teylers Museum, Haarlem

Fragaria vesca

Wild strawberry

In spring, the soil of the shadier parterres in the Rubens Garden is covered with small plants with white flowers, from which edible fruits grow in summer: sweet wild or Alpine strawberries. In Rubens's day, there was a white and a red variant, as documented in Basilius Besler's *Hortus Eystettensis.* They were used against pimples and as blood purifiers, among other things. The strawberries grown for consumption from the mid-eighteenth century onwards were hybrids of American varieties.

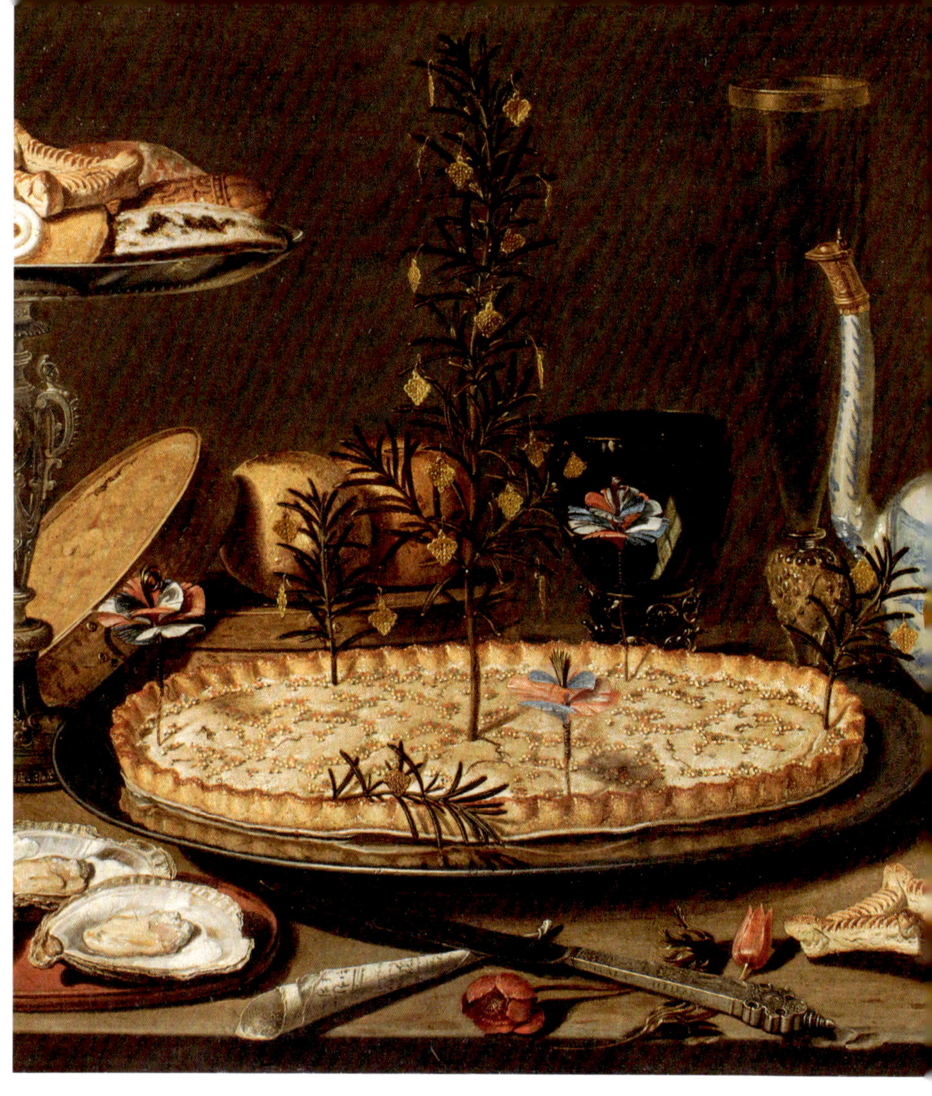

DETAIL FROM Clara Peeters, *Still Life with Tart, Silver Tazza with Sweets, Porcelain, Shells and Oysters*, c. 1612–1613, private collection

According to Dodoens, the fruits had a 'sweet and vinous flavour' and 'did not smell entirely unpleasant'. Lobelius describes how strawberries were eaten as is and were used as table decoration: 'The fruit of this plant, when planted in gardens, is served with other fruit, like raspberries but is more fun to eat'. Wild strawberries are common in early flower and meal still lifes. In some Antwerp paintings from the first quarter of the seventeenth century, rosemary sprigs sometimes appear decorated with *fragulae de auro*, gilded ornaments in the shape of a strawberry, either on the table or standing upright on a cake.

Den verstandighen hovenier over de twaelf maenden van 't jaer of 1672 explains how the juice of wild strawberries can be used to treat pustules and face pimples. Eaten with wine during summer, they are 'invigorating' and purify the blood. Wild strawberry was also used for liver ailments, jaundice, and kidney stones. Two strawberry trees (a shrub and a standard tree) also grow in the Rubens Garden. Like wild strawberry, *Arbutus unedo* was already described by the Roman author Pliny the Elder and was also known in Rubens's time. *Unedo* literally means 'I eat it once'. The fruits were much less popular than the sweet wild strawberries that grew on the ground.

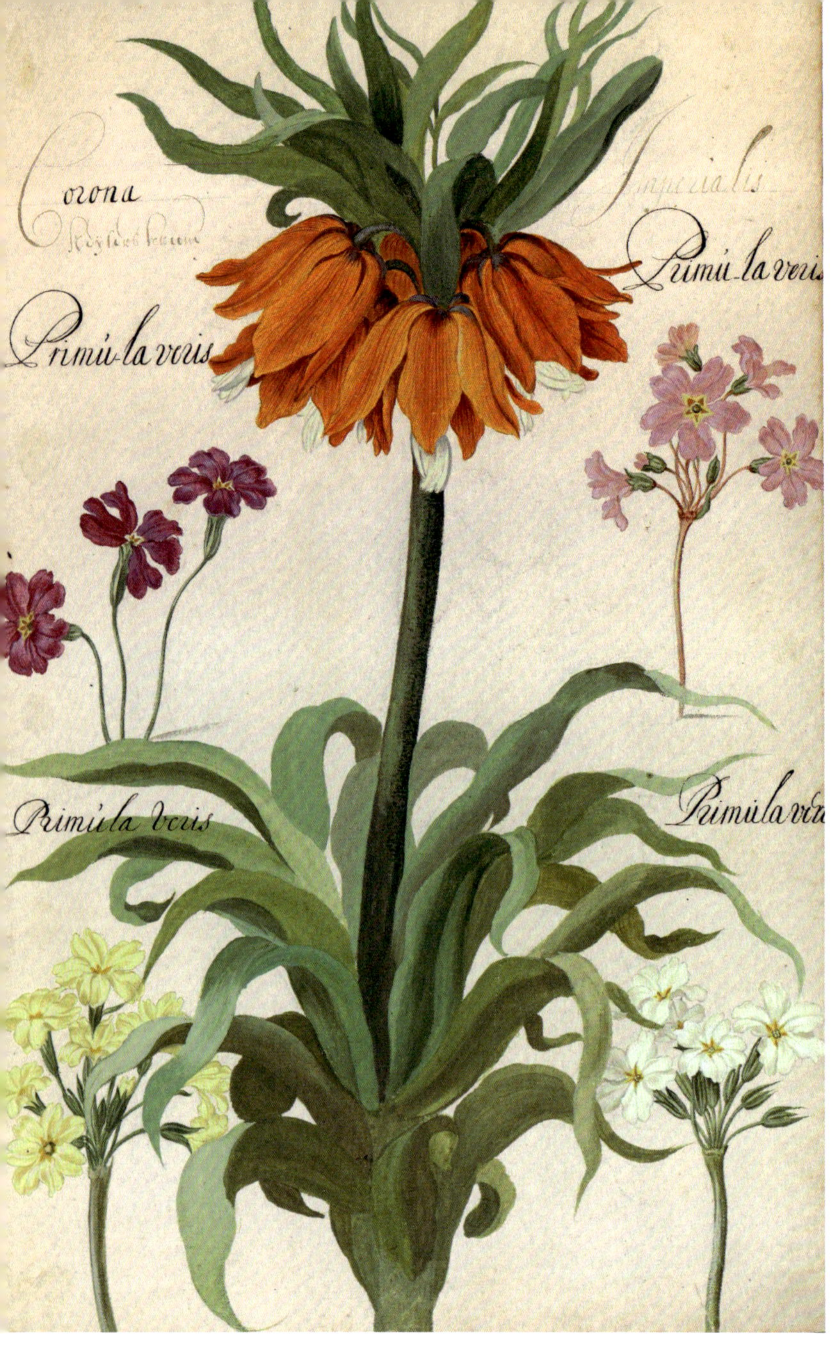

Franciscus de Geest, *Fritillaria imperialis* in the *Hortus Amoenissimus*, 1668, Biblioteca Nazionale Centrale, Rome

Fritillaria imperialis

Crown imperial

In early spring, the *Fritillaria imperialis* or crown imperial towers above all other bulbous plants, scattered throughout the eight large beds of the Rubens Garden. Striking orange bell-shaped flowers are suspended symmetrically and in a perfect circle – like stars around the stem. Contemporary sources do not mention the flower's musky 'fox smell', which is caused by sulphurous terpene. It deters moles, field mice, and voles.

‘In the bottome of ech of these bels there is placed fixe drops of most cleere shining sweete water’, writes English plant expert John Gerard in *The Herball* from 1597, ‘in taste like sugar, resembling in shew faire orient pearles’. The flowers of the fritillary secrete large drops of sweet nectar resembling pearls, much to the delight of butterflies and bees. The glossy and narrow green leaves above the bell-shaped flowers form a crown, a shape that had not been seen on a flower in Europe before.

As in spring gardens, the crown imperial in Baroque flower still lifes often attracts all the attention because of its place at the top of the bouquet. In 1581, Lobelius wrote that this flower was only observed in a few places. He had seen it in Lier at Mr Verdelft’s house, as well as in Marie Schetz’s garden on the Wapper. She had a botanical garden with unusual bulbous plants and trees on the plot next to the later Rubenshuis. In the *Hortus Amoenissimus* , Franciscus de Geest drew a large crown imperial amid four different *Primula* or primroses, which can also be seen in the Rubens Garden. They are among the earliest blooms of the year.

Nor are there many more costly flowers that are not listed and described here, because they are used for decorative purposes in the garden rather than for medicinal uses, such as the beautiful tulips, which one finds in many different colours, and also the beautiful crown imperial. These precious pretty flowers have no nutritional or medicinal uses.

***Den kleynen herbarius ofte kruydt boecxken*, 1640**

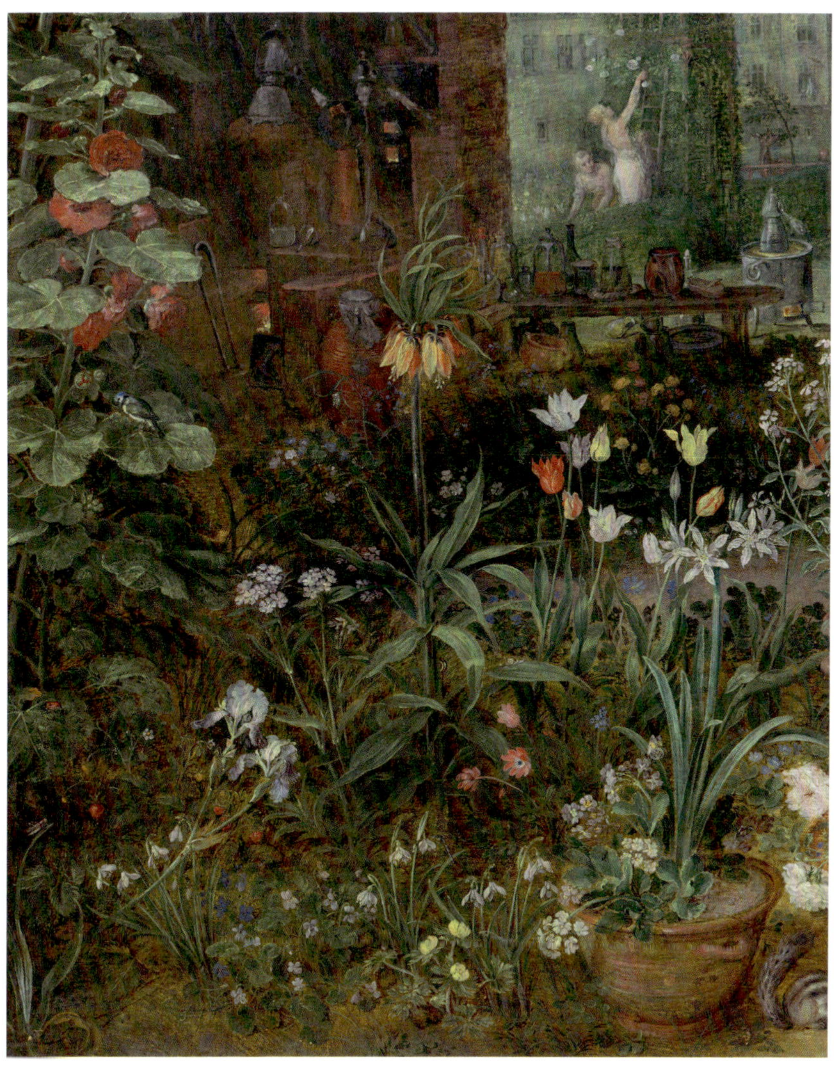

DETAIL FROM Jan Brueghel I and Peter Paul Rubens, *Allegory of Smell*, c. 1617, Museo nacional del Prado, Madrid

Fritillaria meleagris

Snake's head fritillary

In early spring, nearly six hundred purple and white snake's head fritillaries adorn the parterres of the Rubens garden. This elegant stinzen plant is a beautiful native species that can compete with tender tulips, but whose flowers curve downwards, unlike tulips. Snake's head fritillaries combine beauty with humility. They were a popular species in 17th-century gardens and are often depicted in flower still lifes.

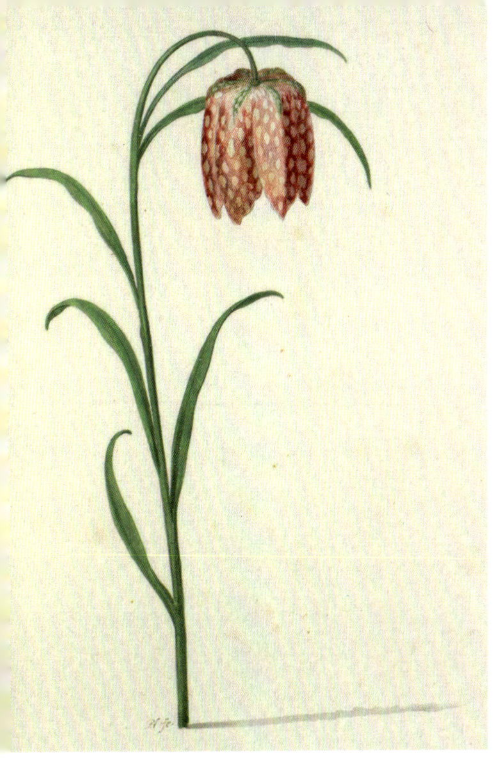

In Lobelius's book of herbs, *Fritillaria meleagris* is meticulously described: 'A thin stem, measuring 30 to 45 cm emerges from the centre of a tiny white bulb, smaller than a daffodil bulb, as if it had already been split into two'. Elongated narrow leaves grown on one, two or occasionally three, pretty flowers curved and arching downwards that are similar to the tulip but smaller, with seven pale yellow stamens, of which the petals are speckled with brown and pale purple spots and stripes. Dodoens recounts how this flower found its way from France to the Netherlands, thriving particularly well there, sometimes with as many as two flowers per stalk. In the garden of a Brussels painter, the rare yellow variety bloomed alongside the purple and white snake's head fritillary in 1609. Franciscus van Sterbeeck described this humble species as *'Kivits-Eyren die uyt ronde bollekens voort-spruyten, de bladen lanck en in den top schoone ghesprenckelde en ghestippelde bloemen dragen tot cieraet van den Hof'* (the eggs of lapwing that spring forth from round spheres, with long leaves and pretty speckled and spotted flowers at the top, embellishing the garden). Pieter Holsteyn II drew a beautiful snake's head fritillary in the 1640s in the flower album *Flores a Petro Holstein ad vivum depicti*, on which the dotted pattern is clearly visible.

Pieter Holsteyn II, *Fritillaria meleagris* in *Flores a Petro Holstein ad vivum depicti*, 1640s, RHS Lindley Collections, London

Humulus lupulus

Hop

One hop plant climbs on the new berceau of the Rubens Garden. The ***Humulus lupulus*** was planted there symbolically, referring to the street Hopland, where there used to be hop fields and vegetable gardens before the advent of bleachfields and urbanisation from the 16th century onwards. Rubens owned several cottages with gardens on the Hopland, adjacent to his plot. You could get to his stables and garden through an informal back entrance on the Hopland. In Rubens's time, hops were used widely and in many ways. Then as now, it was mainly used for brewing beer. But hops were also used as a cooking ingredient and against some ailments.

In Rubens's day, hop plants were already used to cover structures. Dodoens depicts the hop plant in his herbal (1644), writing: 'hops are also eminently suited for covering huts, cottages, arbours, arcades, and flowerbeds'. Hops added shade and a bit of crisp green colour to the garden, 'because it looks like a magnificent green vault from afar, with its many vines and dense foliage'. Hops grow white bell-shaped blooms that are very fragrant.

According to Lobelius, the young hop shoots were eaten as a substitute for lettuce. To make bread lighter and rise faster, you could also add hops to the batter and moisten the dough with the plant's cooking liquid. In 1685, Franciscus van Sterbeeck highlighted how young hop shoots with lettuce and vinegar purified the blood and were beneficial for the liver and spleen. Hops were also said to banish melancholy. Juice of hops relieved earache, and in wine it could help against jaundice and dropsy.

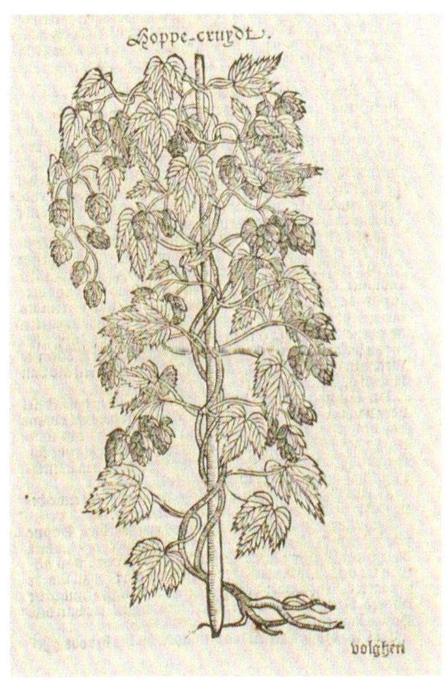

Rembert Dodoens, Carolus Clusius, and Jan Christophe Jegher, *Hoppe-cruydt* in the *Cruydt-boeck Remberti Dodonaei*, 1644, Museum Plantin-Moretus, Antwerp

Jasminum

Jasmine

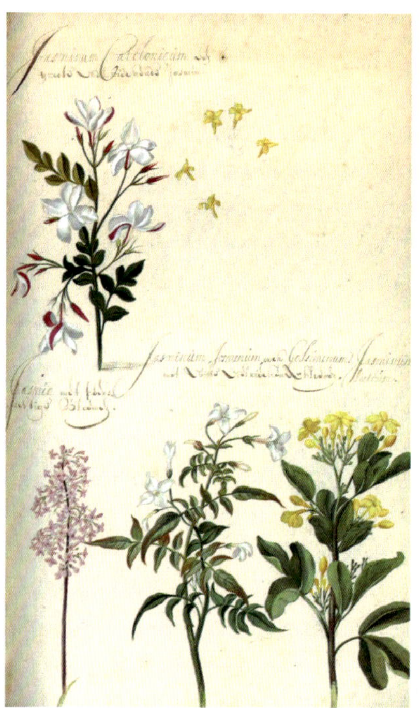

In summer, three fragrant jasmine plants flower in the sunniest spot on the balustrade of the new Rubens Garden. Jasmine was already recommended as a good climber back then and was also grown for its lovely scent. Jasmine generally had white flowers in Rubens's time, but a yellow variety also existed.

In 1581, Lobelius wrote how the plant 'with shoots in the soil, is planted in gardens across England that are well exposed to the sun, and in front of enclosed places in farmyards, and on arbours and berceaus in courtyards, where they are used as an ornamental feature, and trained to grow over an arch with branches'. In *Den Sorgvuldighen Hoevenier* of 1676, jasmine flowers are considered 'praiseworthy' because of their excellent smell. Van Sterbeeck recommended picking them after sunset, as that is when the flowers are driest, most fragrant, and also have the greatest vigour. This also applied to orange blossoms, roses, and daffodils. Not everyone liked the smell of jasmine. Dodoens records in his herbal from 1644 how jasmine was used in public baths but that the scent was so pungent and heavy 'that many don't like to go to the baths as a result'.

Spanish jasmine was most prone to frost and had to overwinter indoors. In early October 1657, citrus as well as oleander, laurel, myrtle, and four *Spaensche jesuminen* grew in the garden of priest and canon Jan van der Veken of Our Lady's Cathedral. Franciscus de Geest depicted both yellow and white jasmine in his *Hortus Amoenissimus*.

LEFT Franciscus de Geest, *Jasminum* in the *Hortus Amoenissimus*, 1668, Biblioteca Nazionale Centrale, Rome

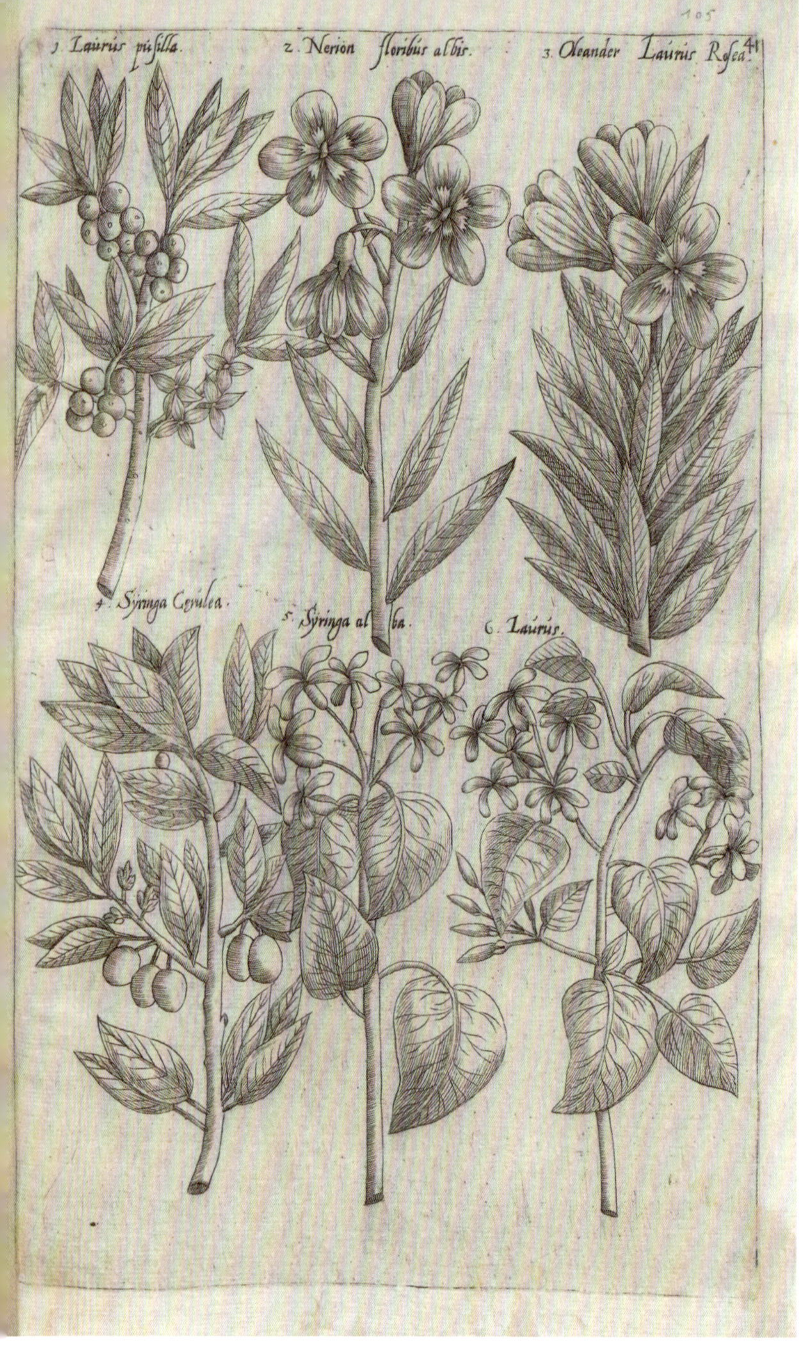

Laurus nobilis in the *Florilegium* by Emanuel Sweerts, 1614, Hendrik Conscience Heritage Library, Antwerp

Laurus nobilis

Bay tree

Six bay trees have been planted in the Rubens Garden: five in the large beds and one in a wooden tub. The evergreen shrub was particularly popular in Rubens's time; there is much evidence of this, including in Antwerp. Bay laurel could be easily propagated by taking cuttings in March. Available in shrubs or trees, bay trees were easy to prune into all kinds of shapes. Many bay trees were frost-sensitive and were therefore often overwintered indoors, especially the young shrubs. A species with somewhat browner leaves that could 'withstand the cold of winter' was already being grown in Bruges at the end of the 16th century.

In January 1620, there were four grafts of bay laurel in the cellar of gardener Jan van Beneden. Baltasar Moretus I paid 21 guilders for 50 of these shrubs on 18 November 1639. In December 1640, there were two bay trees in the cellar the house of Nicolaes Rockox – a friend and patron of Rubens – on Keizerstraat, along with the aforementioned orange trees. In the spring of 1641, the notary recorded a batch of *Lauris cesaris*, a bay tree and four wild laurel shrubs in the home of Alderman Charles de Tassis behind the Capuchin convent. At the end of February 1652, there were citrus plants and two bay trees in the pleasure garden of Antwerp alderman, art collector, and bookseller Jan van Meurs on the Hopland, near Rubens's home. At merchant Olivier van Houte's house on Everdijstraat, there were two bay trees, along with 35 orange and pomegranate trees, in pots, in February 1654. In Margareta Dela Flie's warehouse, you would have found three bay trees in addition to citrus plants in 1655. In early October 1657, there was a *Laurustinus* in the garden of priest and canon Jan van der Veken of Our Lady's Cathedral, a shrub that was the size of a tree. Finally, the extensive bay tree collection of former alderman Joan van Mechelen on Huidevettersstraat is also worth mentioning. In May 1692, there were a striking number of bay trees and shoots in his garden: 17 small and large bay trees, 15 straight bay tree saplings, and 37 pots with bay tree saplings.

In 1676, Reyntkens describes the bay tree as one of the most beautiful trees to grow in the garden in *Den sorgvuldighen hovenier*. According to him, wild laurel was used at all banquets to decorate dishes, and in flower arrangements to sow flowers on them. The French bay tree lent itself particularly well to topiary pruning. According to Dodoens's *Cruydt-boeck* of 1644, laurel was long believed to have special qualities. It could chase away evil spirits and all 'magic', but sprigs in the attic and near windows or doors were also said to keep thunder and lightning away. Fine bay leaves were often used as a cooking ingredient. The black, egg-shaped or almost round berries or *bakelaers* of the bay tree possessed many medicinal properties. They could be used to treat dehydration, poisonous stings, freckles, deafness, stomach, liver, spleen and bladder ailments, skin impurities, and bruises, among other things. A sprig of laurel with berries is depicted in Emanuel Sweerts's *Florilegium* of 1614, which was very popular with gardeners.

TO MAKE SNOW

Take 2.6 litres of cream and the whites of 10 or 12 eggs, beaten until stiff, some rosewater and sugar, and beat it well until you obtain foam. You can spoon this snow into platters or on rosemary or bay trees or branches.

***Antwerps kookboek van Clara van Molle*, first half of the 17th century**

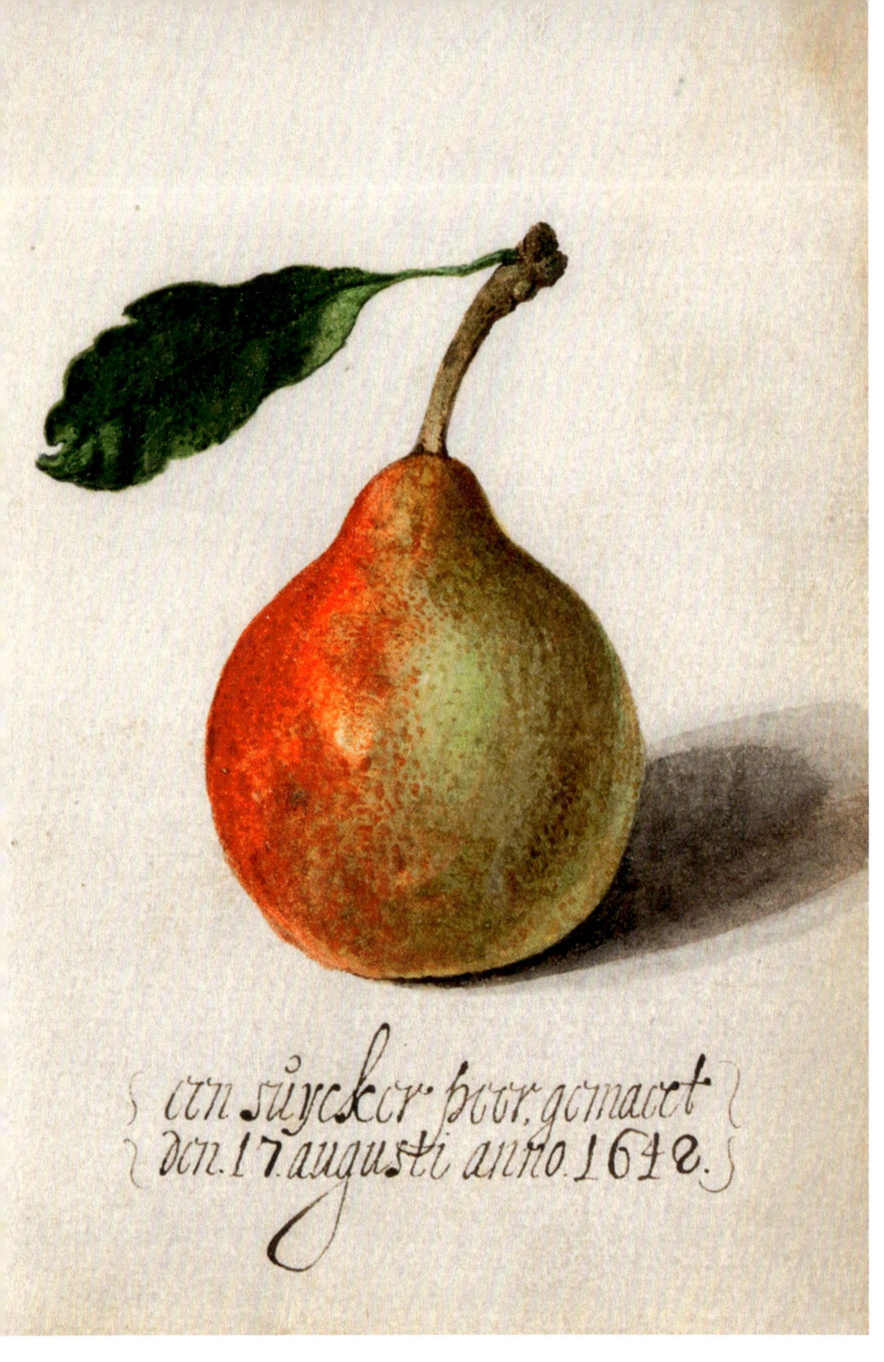

Pieter Serwouters, *Suikerpeer (Sugar Pear)*, 17 August 1648,
Drents Museum, Assen

Pyrus 'Rousselet de Reims'

Pear

Besides the fig tree, there is also a stunning pear tree in the parterre in front of the pavilion. Together with the pear, orange, and lime trees, this is the only tree of which we can be really certain that it once stood in Rubens's garden. But for now, the search for a 17th-century variety named *Rosile* that is ripe in mid-August (Rubens mentions this pear variety in a note dated 17 August 1638) continues. During the 2024 redevelopment, a *Rousselet de Reims* was planted in Rubens's garden. Etymologically speaking, this could be the *Rosile* that Rubens refers to in his note to Lucas Faydherbe.

Proof of the presence of a pear tree in Rubens's garden can be found in the already quoted note that Rubens sent from Elewijt to his friend and caretaker Lucas Faydherbe in Antwerp. In it, Rubens asked Faydherbe to remind Willem, the gardener, of the *Rosilepeyrkens* (Rosile pears) and figs, should there be any, and any other delicacies from his garden.

The French agronomist Olivier de Serres mentions a *de Rousselet* pear in his *Le théâtre d'agriculture et mesnage des champs* from 1600, a book for gardeners that could also be found in Nicolaes Rockox's library. *Le jardinier solitaire* of 1612 describes the *Rousselet de Reims* as one of the best summer pears available, *demi beurré* with *beurré* meaning butter-like in reference to the juicy flesh of pears, and of different sizes depending on the pear tree's growth habit – *en espalier* or *en plein vent* (espaliered or growing freely). Little *Rousselet* is praised for its taste, which is *plus relevé* (pronounced). The pear is said to keep longer and is extremely suitable for making confit. The book also lists the *Robine*, a summer pear that ripens in July and August: *'La Robine qui se nomme aussi la Royale d'esté, est petite (...) son fruit vient sur les arbres par bouquets; elle est tres-musquée, sucrée, & estimée des curieux'* (*Robine* or *Royale d'esté* is small and grows in bunches on the tree. It has a very musky, sweet flavour and is widely appreciated by curious palates). Besides the old pomologies (reference works listing all the fruit varieties of the time), painted pears in Antwerp fruit still lifes by Rubens's contemporaries and colleagues such as Osias Beert may also help us determine which pear variety Rubens was referring to.

Lesquels noms par la revolution des siecles, ne sont plus cognues aujourd'hui, non plus que plusieurs autres imposés aux fruits pour diverses causes, prinses, ou des lieux d'où premierement ont esté tirés, ou de leurs marques particuliers, ou du temps de leur maturité, & semblables qui ont confondu leurs appellations. Et quant mesme en ce temps-ci demeureroient telles, ne les pouvons-nous pourtant rapporter à nos Poires, pour icelles appellations n'avoir esté pareilles generalement par tout: à quoi de necessité eust fallu que tous peuples eussent consenti, si on les eust voulu conserver, & sans confusion s'en servir. Mais tout le rebours s'est pratiqué depuis plusieurs siecles, car non seulement chacune province, ains presque chaque privé enteur, a donné à ses fruits, noms selon sa fantasie.

The names that have fallen into oblivion over the centuries, like many others given to fruits for various reasons – because of where they originally came from or because of their characteristics or because of the period of ripeness, and so on – are confusing. Even if these names still existed in this day and age, we could not use them for our pears, because these designations were not universally the same: it was necessary for everyone to agree, if they had wanted to keep them and use them without confusion. But the opposite has been the case for centuries, as every province and almost every individual grower has named his fruit according to his own imagination.

Olivier de Serres, *Le théâtre d'agriculture et mesnage des champs*, 1600

APPLE OR PEAR TART

Use sweet pears. Peel and core them. If they are small, slice them into small rounds. If they are large, slice them into small, thin slices. Prepare the base. Combine flour, two egg yolks and fresh butter with some water. Roll out the dough until thin. Add some saffron to the dough for colour. Then add the pears with fresh butter or small chunks of bone marrow, currants, cinnamon, some ginger powder and powdered sugar. Cover it with a lid made of dough and bake it in the oven or pan. Brush with butter when done, and sprinkle over some cinnamon or grated sugar. They also make apple tarts this way, but with fennel seeds.

***Antwerps kookboek van Clara van Molle*, first half of the 17th century**

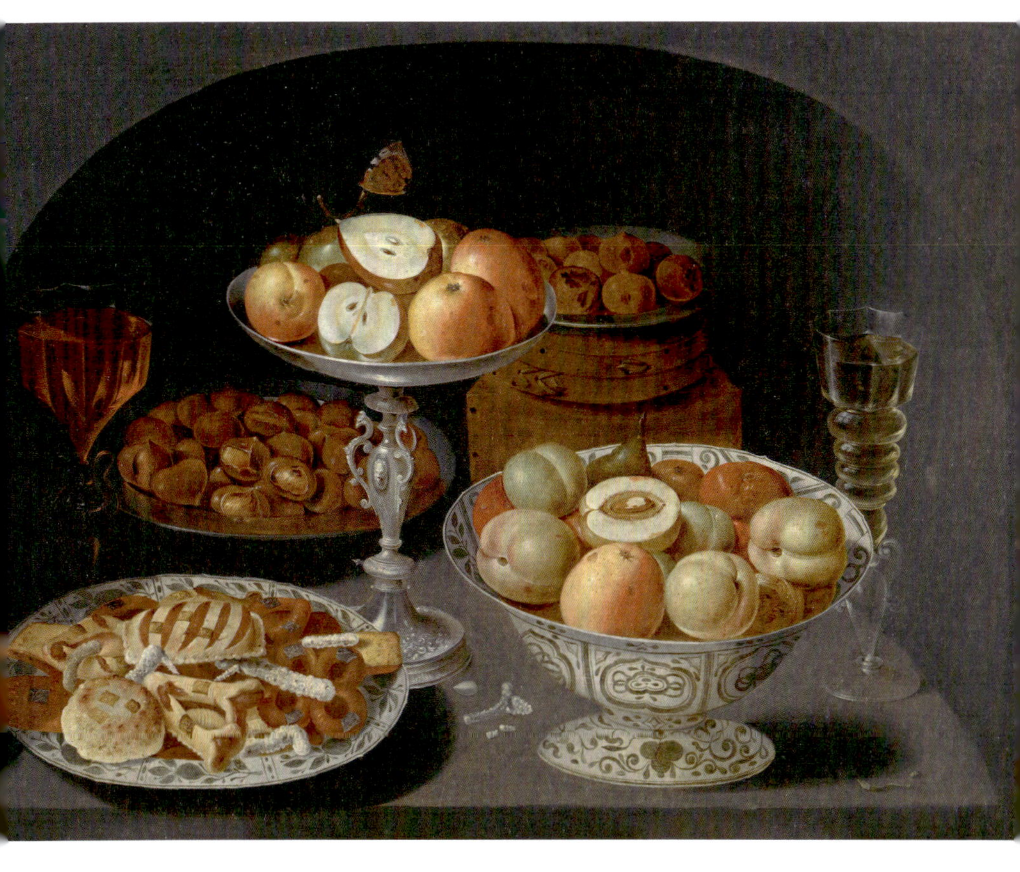

Osias Beert, *Fruit, Confectionery and Wine in a Niche*, c. 1615–1620, private collection

Rosmarinus

Rosemary

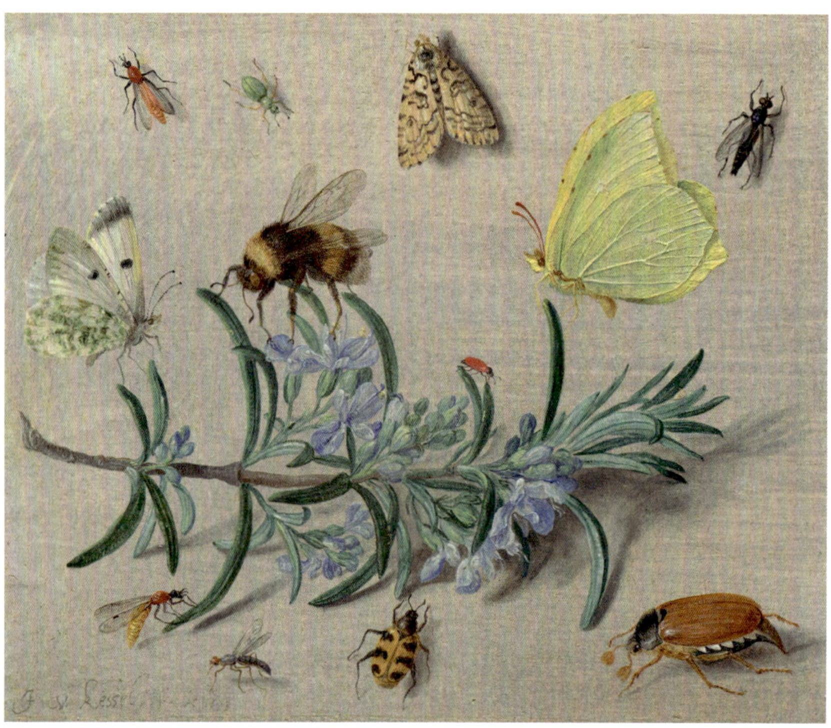

***Rosmarinus prostratus* 'Repens', the creeping and richly flowering variety of the well-known kitchen herb, blooms in early spring on the outer sides of the parterres. Rosemary already had many medicinal uses in Rubens's time.**

Jan van Kessel I, *Insects and a Sprig of Rosemary*, 1653, Courtesy National Gallery of Art, Washington

In his *Le théâtre d'agriculture et mesnage des champs* from 1600, Olivier de Serres recommends fragrant rosemary, sage, marjoram, and lavender, among others, as border plants for geometric garden beds. Dodoens writes the following about this in his herbal: 'In Italy, they plant their garden beds with rosemary, but here it is planted and maintained with great care as a decorative feature for gardens'. Rosemary was also well-suited for lining a berceau, along with shrubs whose branches could be easily trained and tied back, such as roses, jasmine, ivy, myrtle, hops, and woodland vine. Perhaps this was also the case for the three rosemary trees in late February 1652 that could be found in the pleasure garden of Antwerp alderman, art collector, and bookseller Jan van Meurs, who lived near Rubens on the Hopland. Small rosemary trees were not frost-resistant and overwintered indoors in Rubens's time. Those who had no space for them indoors, could cluster them together on a table in the garden, making sure to wrap them well in case of frost to prevent them from freezing, according to *Den sorgvuldighen hovenier* of 1676.

Rosemary had all kinds of medicinal uses and was widely used as a home remedy. When steeped in wine, it could help with heart problems, concentration disorders, and a diseased liver and spleen. Rosemary was also used as a remedy for forgetfulness. For good breath, you could chew rosemary. Rosemary juice with honey helped with dark eyes and refreshed the face, according to the Antwerp edition of the *Den verstandighen hovenier*.

Dodoens had already described how the many thin branches were covered with elongated, narrow, and hard leaves, which smelt nice. 'Tiny flowers that are blue verging on white grow from its roots or foliage', according to his *Cruydt-boeck* of 1644. This can also be seen in a small panel painting by the Antwerp artist Jan van Kessel from 1653.

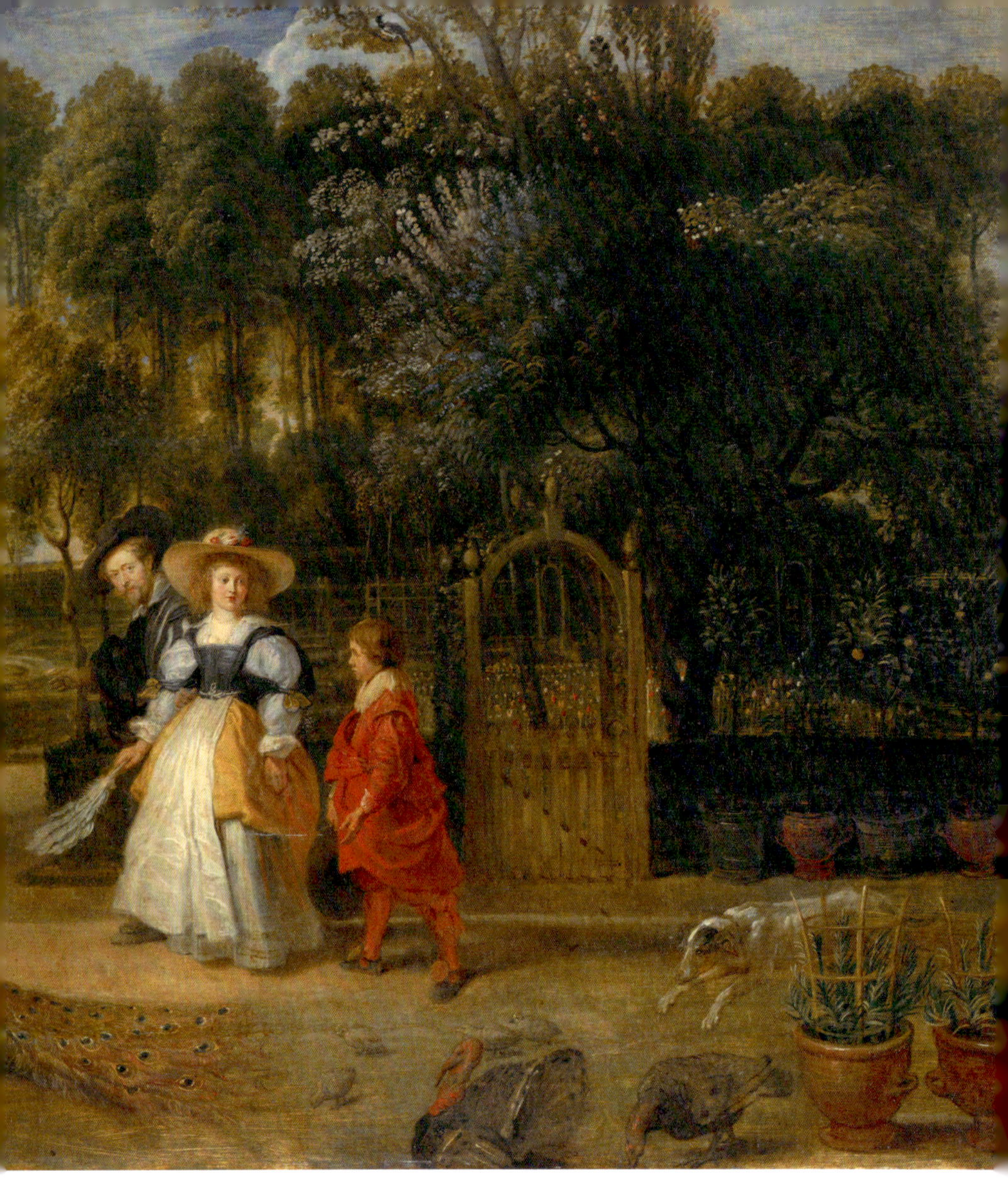

DETAIL FROM Peter Paul Rubens (atelier), *Peter Paul Rubens, Helena Fourment and Their Son Nicolaas Walking in Their Garden ('The Walk in the Garden')*, c. 1630–1631, Bayerische Staatsgemäldesammlungen – Alte Pinakothek, Munich

Taxus baccata

Yew

The stately giant yew that towers over the portico is one of the eye-catchers in the Rubens Garden. The yew tree may be somewhat younger than Rubens, but trees have always been a defining feature of the garden, even long after Rubens's death and the sale of the house. Originally, a second specimen stood there, as shown in old photos. During the reconstruction of the museum and the redevelopment of the garden in the 1940s, it appeared to have died. Nearly 450 yew hedges also line the outer perimeter of the parterres in the new Rubens Garden.

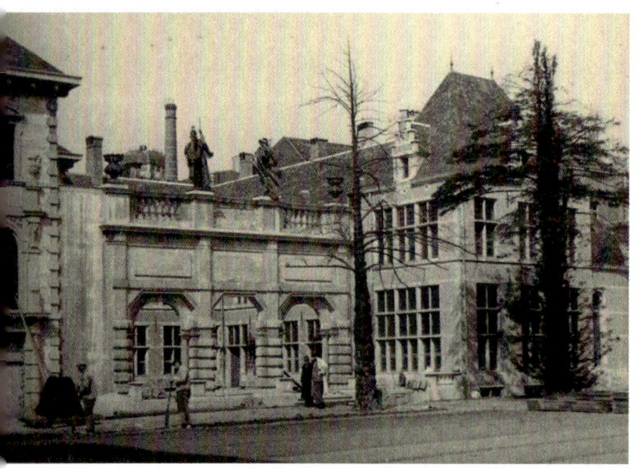

According to an urban legend, these yew trees were planted by Rubens himself, to mark the birth of his two eldest sons, Albert and Nicolaas. However, research showed that the yew in the Rubens Garden may be 'only' 250 years old. Together with its felled counterpart, it was presumably planted by the de Bosschaert de Pret family, which acquired the house in 1763, modernising it extensively between 1766 and 1791 in a late 18th-century classicist style and with a completely redesigned landscape garden. Initially, the plan was to preserve the second yew tree at all costs when the garden was redesigned in the 1940s and have it overgrown with ivy or plants with small leaves. But below the surface, nature was already taking its course: the roots turned out to be infested with worms, that had already wreaked too much damage by the time they were discovered. As a result, the second yew also had to be felled.

As in the painting *The Walk in the Garden*, the corners of today's yew hedges are not straight but rounded. Slate shingles of sufficient thickness were reused as a border for the hedges. They come from the first evocation of the garden from 1946, when the design consisted of four large beds with low yew hedges, each accommodating four grassy areas bordered with boxwood. Even then, the slate tiles had already been reclaimed. *The Walk in the Garden* features a remarkable number of trees of all shapes and sizes. Shortly after Rubens's death, the artist's cousin, Philip, explained in the *Vita Petri Pauli Rubenii* how his uncle had bought a house with land in Antwerp, where he built a large house in the Roman

View from the garden with two yew trees during the restoration of the Rubenshuis, 1940s, Rubenshuis, Antwerp

style. The house had been remodelled so that he could use it as a studio, and it also had a spacious garden with all kinds of trees (in Latin: *omnis generis arboribus*).

Trees are mentioned in almost every deed of sale from 1755 onwards, each time the house and garden changed hands. An advertisement for the public sale of a beautiful large house with a garden was published in *Gazet van Antwerpen*, the local newspaper, on 26 December 1755. It describes various elements of the garden: 'garden statues, pedestals, copper vases in the garden, etc.'. All climbers, hedges, and palms were also included in the sale. The deed of sale of the Rubenshuis to the Bosschaert family in 1763 also mentions figures and pedestals, a copper vase standing at the end of the garden, with hedges, trees, and benefits in that same garden.

The current yew guards many newly-planted trees that also ensure the garden is green in winter: *Cydonia oblonga 'Champion'* (quince), *Ficus carica* (fig), *Gymnocladus dioica* (Kentucky coffee tree), *Koelreuteria paniculata* (Chinese lacquer tree), *Magnolia grandiflora 'Ferruginea'* (evergreen magnolia), *Malus domestica 'Red Autumn Calville'* (raspberry apple), *Mespilus germanica* (medlar), *Morus alba* (white mulberry), *Prunus persica 'Reine des Vergers'* (peach), *Pyrus communis 'Rousselet de Reims'* (pear), *Quercus nigra* (water oak), *Quercus x turneri 'Pseudoturneri'* (Turner's oak) and *Robinia pseudoacacia* (false acacia).

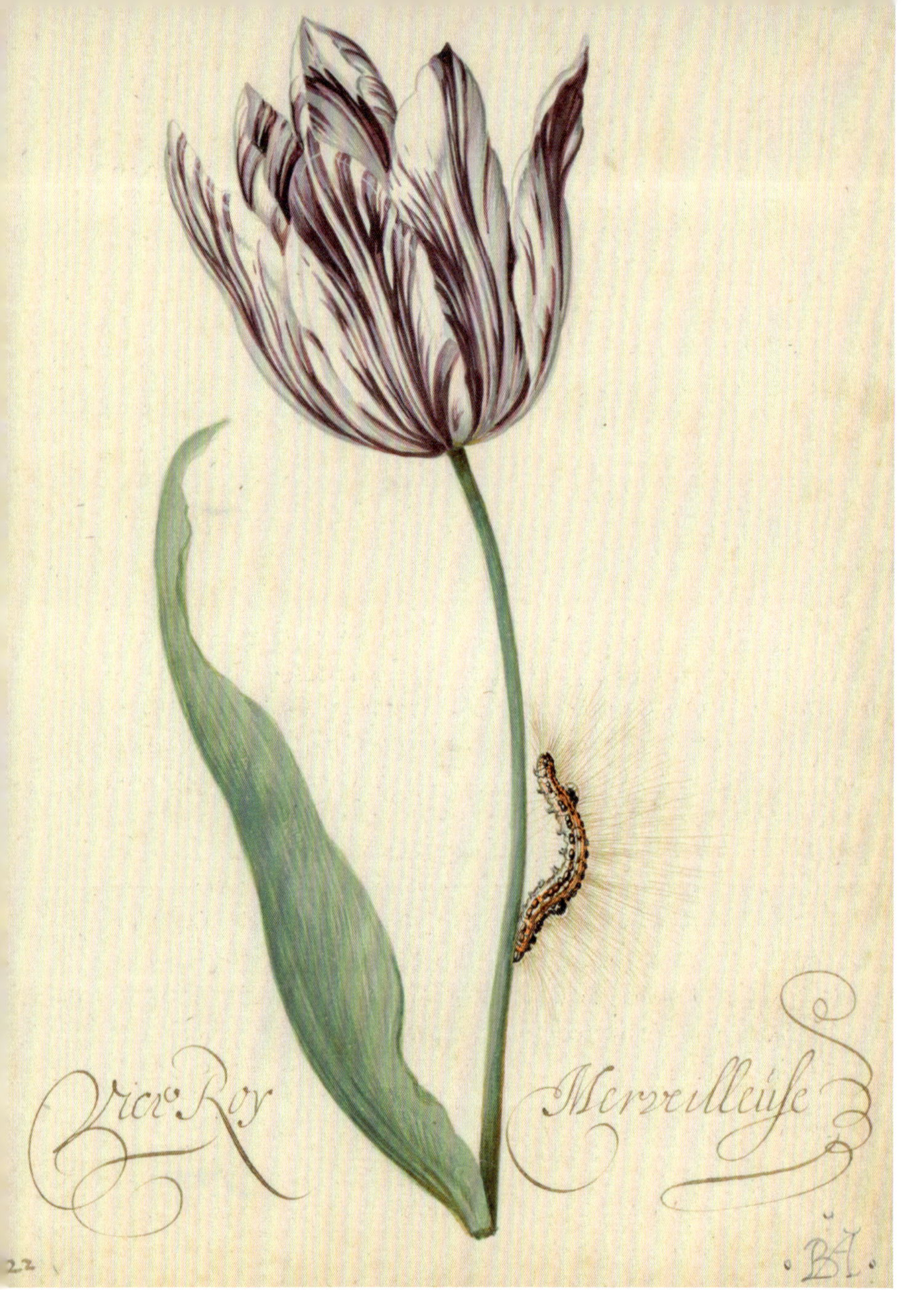

Baltasar van der Ast, *Viceroy Merveilleuse*, 1632–1657, Fondation Custodia – Collection Frits Lugt, Paris

Tulipa

Tulip

In spring, just under a thousand tulips, in all shades of yellow, red, and purple, bring colour to the parterres of the new museum garden. A total of 22 varieties were planted, divided by colour in different 'tulip pockets': *Absalon, Acuminata, Alba Regalis, Bacchus, Batalinii 'Bronze Charm', 'Black and White', Clusiana, Clusiana 'peppermintstick', Duc van tol Red and Yellow, Duc van tol Violet, Eichleri, Insulinde, Julia Farnese, Lac van Rijn, Little Princes, Maryland, Rubens, Sprengeri, Sylvestris, The Lizard, Turkestanica,* and *Whittallii*. Tulips are by far the most frequently mentioned flower in archival sources about Antwerp city gardens in Rubens's time. New archival research for the Rubens Garden also revealed a novel fact, namely that there also was an Antwerp tulip mania. Rubens maintained contacts with some tulip speculators. In 1637, the tulip bubble burst, wreaking financial havoc.

As in the Northern Netherlands, speculation drove the value of tulip bulbs to extremes in Antwerp in the 1630s. A small group of collectors organised meetings at florists' shops and in collectors' homes, as well as at inns. In Antwerp, they met in *De Zwaan*, the brewery and inn of Hendrik Stockmans at the Brabantse Korenmarkt – the present-day Graanmarkt. Bulbs were sold over (or under) the counter, and a bulb might even have changed hands several times in the course of an evening. The price was determined by the rarity of the tulip as well as the bulb's weight. In St George's Church, a stone's throw from Rubens's home and garden, tulip collectors held masses so that tulips would grow well, and a statue of Flora was even installed there, adorned with roses and floral wreaths. After the mass, tulip speculators would gather at the inn to strengthen their ties. An altar and brotherhood to Saint Dorothy, the patron saint of florists and gardeners,

En sorte que la vente & revente publique en estant interdite, on en est venu aux eschanges & ventes particulieres. Mais comme cela ne se peut sans qu'il arrive quelquefois du trouble entre les Hommes, les curieux Flamans ont institué par les Villes une Confrairie, pour laquelle ils ont pris saincte Dorotée pour Patronne.

Because public buying and selling [of tulip bulbs] was banned, people turned to private exchanges and sales. But as this cannot be done without a dispute every now and then, the curious Flemings established a brotherhood in their town, for which they chose Saint Dorothy as their patron saint.

Charles de la Chesnée de Monstereul, *Le floriste françois*, 1653

were even founded in St George's Church on their initiative. A play was dedicated to the saint in 1641, presumably to mark the brotherhood's founding. In the song for Dorothy, *Tulpans aller deughden* (tulips of all virtues) are the *oorspronck aller vreughden* (origin of all joy). The play was dedicated to Jacques van Eyck, head of a wealthy merchant family. Jacques's son Jacobus married Cornelia Hillewerve. The couple became the new owners of the Rubenshuis in 1660.

Tulip bulb prices peaked from 1637 onwards, by which time many buyers could no longer pay for their bulbs. In June, Antonio de Tassis wrote from Antwerp that he had sold a *Viceroy* tulip – an early tulip, white and purple – to brewer Hendrik Stockmans for 1,800 guilders, but that Stockmans had failed to pay for it. The *Viceroy* urgently needed to be paid for and collected, but the

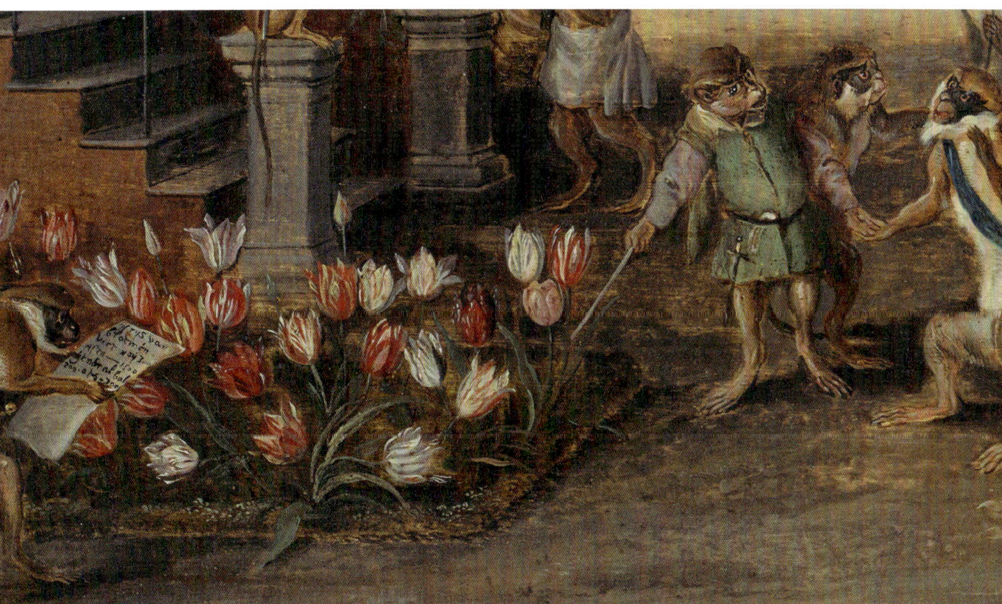

DETAIL FROM Jan Brueghel II, *Satire of Tulipomania*, c. 1637, Frans Hals Museum, Haarlem

notary thought there was very little chance of this now that tulip prices were being slashed. Many growers and collectors went bankrupt and were left with bulbs, resulting in sometimes long notarised testimonials. More than five hundred varieties circulated in the Netherlands at the time. Thanks to newly-discovered documents, we now know exactly which tulips flowered and were traded in Antwerp in Rubens's time. Some species refer to a grower: *Van Dyck*, *Van Kaer*, *Van Eycken*, *Catlijn*, *Jan Geertssen*, and *De Man*. Others refer to the city where they were (first) cultivated: *Oudenaarde*, *Brandenburg*, *Tournai*, *Tourlon*, and *Gouda*. *Admiraals* and *Generaals* were tulips of great beauty that had won competitions. Other tulips that were featured in Antwerp deeds include *Petters*, *Switers*, *Harmoyants*, *Pereltkens*, *Agaten robijn*, *Morlion*, and the *Somerschoon*.

The discovery of an Antwerp tulip bulb club is significant in the research into the Rubens Garden, as there are direct links between Rubens and some tulip speculators. Antonio de Tassis, a pivotal figure, commissioned a portrait of himself and his daughter from Rubens's apprentice Anthony van Dyck, and also owned *The Massacre of the Innocents* by Rubens for some time. De Tassis had a large garden on the Stijfselrui, the street behind the Capuchin convent. He kept his own inventory, consisting of a 25-page plant book. Another member of the group of tulip enthusiasts was Peter Hannekart. From 1637, Hannekart owned a property on the Oudaan with a large garden. He also had six large paintings by Rubens, including the unfinished *Henry IV at the Battle of Ivry*, which is now in the collection of the Rubenshuis. Hannekart and Rubens were brothers-in-law. After Rubens's death, he became the *co-momboir* or guardian of the five minor children of the artist and Helena Fourment. Tulip speculators Gaspar Charles and Baltasar van Engelen were also part of Rubens's network of clients in Antwerp. The speculator

Hendrick Stockmans, who was also a brewer and the owner of *De Zwaan* brewery and inn, had a beautiful flower garden with many tulips. We know from the settlement of the estate of Isabella Brant, Rubens's first wife, that Stockmans delivered beer to the Rubens family. Did the delivery sometimes also include the odd tulip bulb? Anyone who looks closely at the plantings in the parterre in the painting *The Walk in the Garden* will immediately spot the colourful tulip garden where *Goudas*, *Viceroys*, *Van Dycks*, and *Somerschoon* flourished to everyone's delight.

For a few years, these flowers were very valuable, especially those with rare and magnificent blooms, because people paid a lot of money for a single flower, the visual splendour of which they could enjoy for just a brief time, without any other amusement, profit or advantage, whether in food or medicinal.

FVS, *Den verstandighen hovenier*, 1672

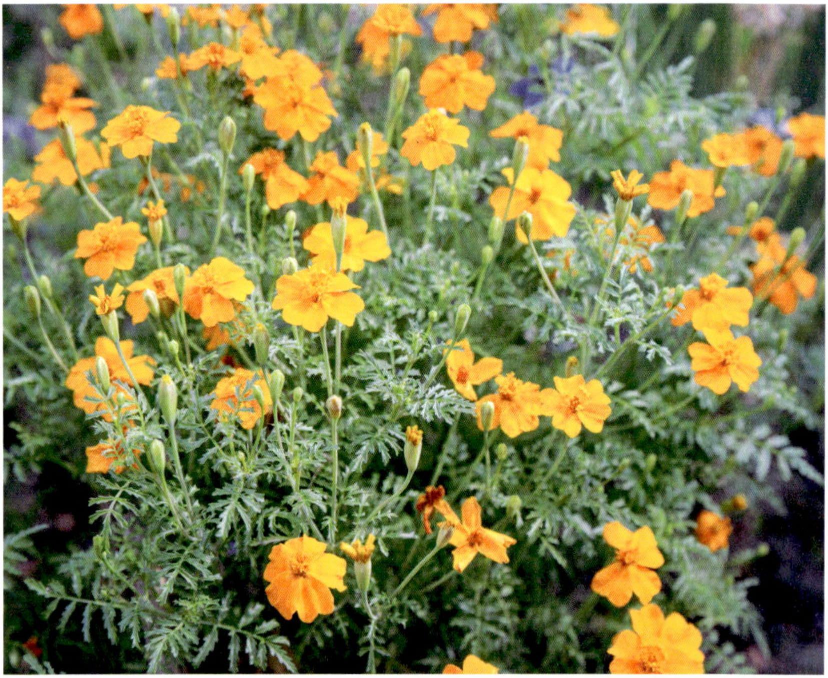

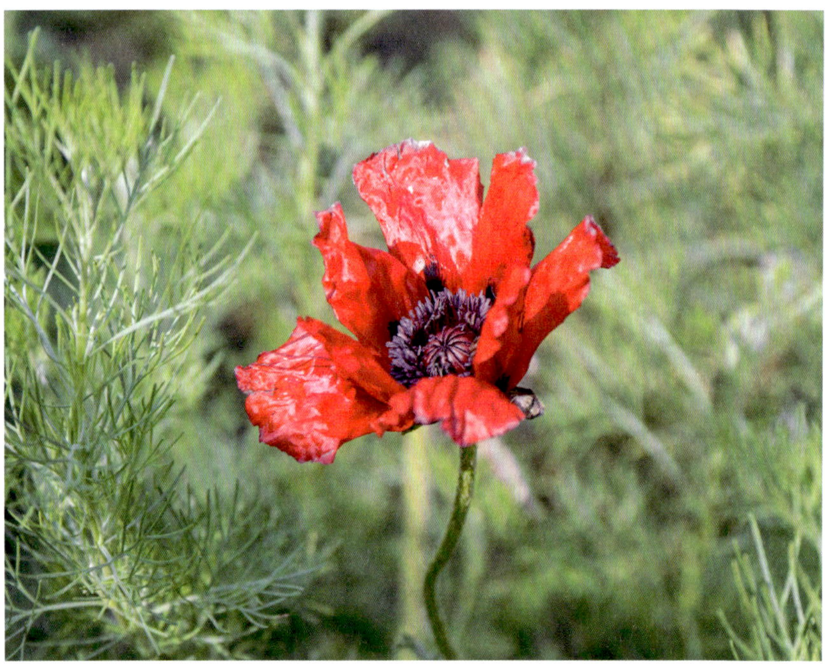

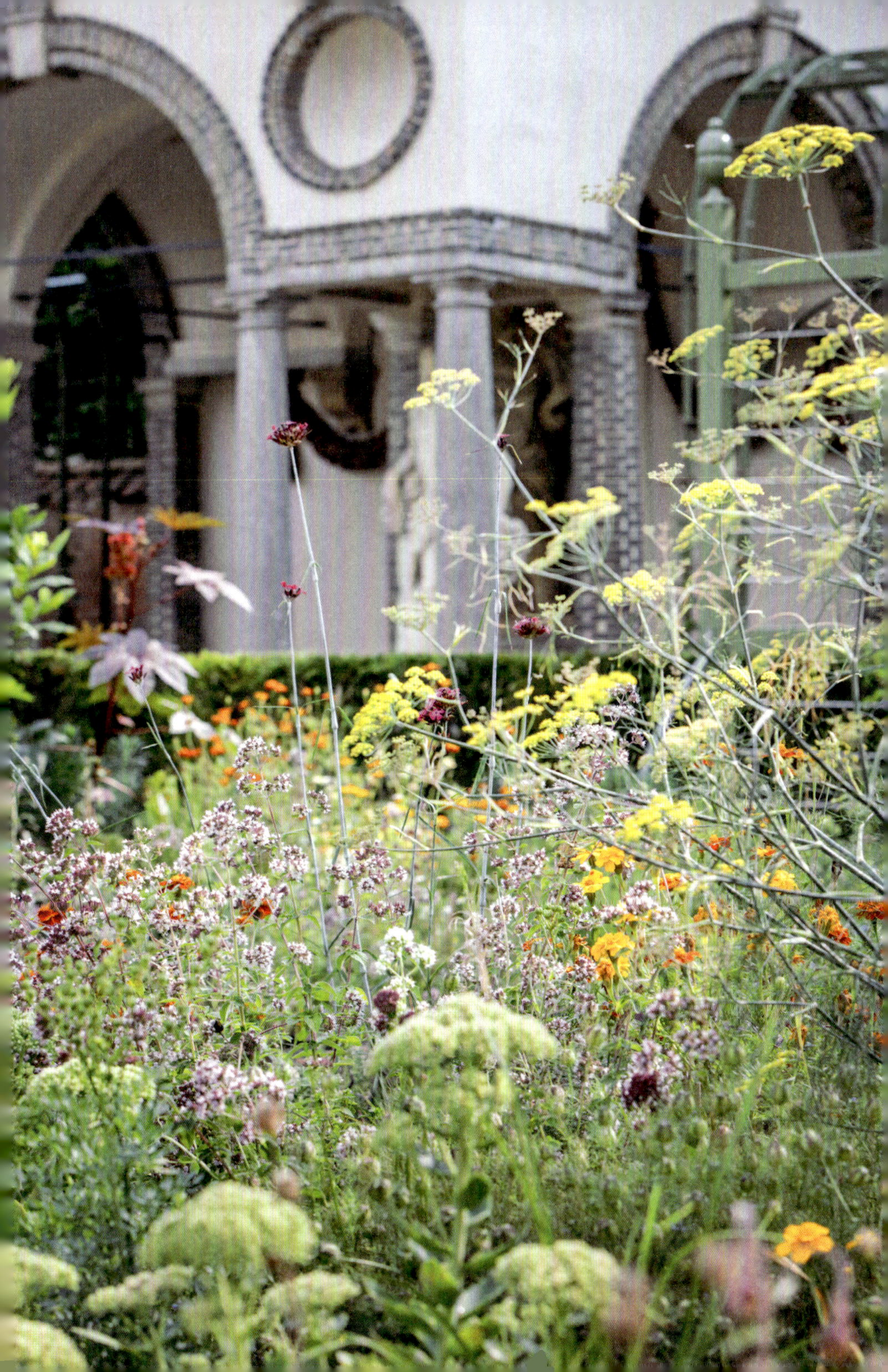

Further reading

Alen, Klara, *Oft iets anders treffelijck uit den hoff: een verkennend archief- en literatuuronderzoek voor een historisch verantwoorde beplanting en een museale omkadering van de Rubenstuin*, unpublished report, 2021.

Alen, Klara, 'Met vergrootglas en penseel. Bloemen en bloemstukken in de Nederlanden tussen 1560 en 1620', *Power Flower. Bloemstillevens in de Nederlanden*, exh. cat., Rockoxhuis, Antwerp 2015–2016, pp. 7–33.

De Backer, Walter, Paul Huvenne and Herman Van Den Bossche, 'De tuin van het Rubenshuis in Antwerpen: een historische evocatie', *Monumenten, landschappen en archeologie*, 14, 3 (1995), pp. 41–56.

Büttner, Nils, *Herr P.P. Rubens: von der Kunst, berühmt zu werden*, Göttingen 2006.

Van Driessche, Thomas, '"Ten einde de historische werkelijkheid op een zo volmaakt mogelijke wijze te benaderen". Emiel Van Averbeke en de "herschepping" van het Rubenshuis', *Monumenten, landschappen en archeologie*, 32 (2013), pp. 12–43.

Gärten und Höfe der Rubenszeit im Spiegel der Malerfamilie Bruegel und der Künstler um Peter Paul Rubens, exh. cat., Städtisches Gustav-Lübcke-Museum, Hamm (Westfalen) 2000–2001.

Heinen, Ulrich, 'Rubens' Garten und die Gesundheit des Künstlers', *Wallraf-Richartz-Jahrbuch*, 65 (2004), pp. 71–182.

Kendal, Anne, 'The Garden of the Rubens House, Antwerp', *Garden History*, 5, 2 (1977), pp. 27–29.

Maclot, Petra, 'De Rubenssite: van pre-industrieel complex tot museale kunstenaarswoning', *Monumenten, landschappen en archeologie*, 39 (2019), pp. 20–33.

Maclot, Petra, *Bouwhistorisch onderzoek, analyse en waardenstelling van de Rubenssite: het als monument beschermde Rubenshuis, Wapper 9-15 en Hopland 13, en Kolveniershof met Rubenianum, Kolvenierstraat 16-20, 2000 Antwerpen (Antwerpen, afdeling 3, Sectie C 1358f en 1370e)*, unpublished report, 2016.

De Maegd, Chris, 'Les jardins sont la partie la plus riante d'une maison. Siertuinen bij adellijke residenties in de 17de en de 18de eeuw', *Revue belge de philologie et d'histoire. Histoire médiévale moderne et contemporaine*, 88, 2 (2010), pp. 409–434.

de Nave, Francine and Dirk Imhof (ed.), *De botanica in de Zuidelijke Nederlanden (einde 15de eeuw–c. 1650)*, exh. cat., Museum Plantin-Moretus, Antwerp 1993.

De Poorter, Nora and Frans Baudouin, *Rubens's House*, Corpus Rubenianum Ludwig Burchard 22, 2, London 2022.

Segal, Sam and Klara Alen, *Dutch and Flemish Flower Pieces. Paintings, Drawings and Prints up to the Nineteenth Century*, Leiden 2020.

Uppenkamp, Barbara and Ben Van Beneden, *Palazzo Rubens: de meester als architect*, exh. cat., Rubenshuis, Antwerp 2011.

Van de Velde, Hildegard, *Antwerpen, een originele stadstuin rijker. Thien oraignie boomen, sommigen met appelen belaeden*, exh. cat., Rockoxhuis, Antwerp 2002.

Watteeuw, Bert, '"Hy weet wonderlijcke dinghen te weghe te brenghen." Snoeisel en maaisel uit de zeventiende-eeuwse Antwerpse tuincultuur', *Power Flower. Bloemstillevens in de Nederlanden*, exh. cat., Rockoxhuis, Antwerp 2015–2016, pp. 95–113.

Watteeuw, Bert, 'Onzichtbaar aanwezig: dienstpersoneel ten huize van Rubens', *Rubens privé. De meester portretteert zijn familie*, exh. cat., Rubenshuis, Antwerp 2015, pp. 54–75.

Photo credits

The Rubens Garden in 2024. © Ans Brys (p. 4, 8, 36, 39, 45, 53, 73, 154, 155).

Adriaen Collaert, *Dianthus* in the *Florilegium*, c. 1590, Rijksmuseum, Amsterdam.

Archaeological finds, City of Antwerp Archaeology Department, 2024. Photo: Frederik Beyens.

Baltasar van der Ast, *Lemoen met haer Bloessem*, 1632–1657, private collection.

Baltasar van der Ast, *Viceroy Merveilleuse*, 1632–1657, Fondation Custodia – Collection Frits Lugt, Paris.

Basilius Besler, *Citri, Cyclamen hederifolium* and *Fragaria vesca*, 1613, Teylers Museum, Haarlem.

Bernard Wynhouts, *Erythronium dens-canis* in the *Herbarius continens species plantarum tum patriarum tum exoticarum ad vivum prout nascuntur in horto infirmariae celeberrimae Abbatiae Diligemensis*, 1633, University Library, Ghent.

Charles Emmanuel Bizet and Franciscus Ertinger, Title page for Franciscus van Sterbeeck's *Citricultura*, 1682, G 10427, Collection of the City of Antwerp, Hendrik Conscience Heritage Library.

Clara Peeters, *Flowers in a Stoneware Vase, with a Pot with Carnations*, 1612, private collection.

Clara Peeters, *Still Life with Tart, Silver Tazza with Sweets, Porcelain, Shells and Oysters*, c. 1612–1613, private collection.

Copper plate of *Cyclamen hederifolium* in preparation for the engraving in the *Hortus Eystettenis*, Albertina, Vienna.

Crispijn de Passe II, *Spring, Narcissus, Anemone* and *Hepatica* from the *Hortus Floridus*, 1614, Oak Spring Garden Foundation, Upperville, Virginia (USA).

David Teniers II, *An Elegant Company Before a Pavilion in an Ornamental Garden*, 1651, © The Phoebus Foundation, Antwerp.

David Teniers II, *Spring*, c. 1644, The National Gallery, London, NG857, acquired, 1871.

Drawing of *Cyclamen hederifolium* in preparation for the engraving in the *Hortus Eystettenis*, Universitätsbibliothek, Erlangen-Nuremberg.

Emanuel Sweerts, *Aquilegia vulgaris* and *Laurus nobilis*, 1614, G 88824, Collection of the City of Antwerp, Hendrik Conscience Heritage Library.

Franciscus de Geest, *Borago, Calendula, Cyclamen, Dianthus, Fritillaria imperialis* and *Jasminum* in the *Hortus Amoenissimus*, 1668, Biblioteca Nazionale Centrale 'Vittorio Emanuele II' di Roma.

Garden pavilion before the restoration of the Rubenshuis, early 20th century, PHOTO-OF#1173, Felixarchief, Antwerp.

Inventory of the estate of Alderman Charles de Tassis, N 2285, Felixarchief, Antwerp.

Jacob Harrewijn after Jacob van Croes, *The Rubens House at Antwerp ('Parties de La Maison Hilwerve à Anvers')*, 1692, Collection of the City of Antwerp, Rubenshuis. Photo: Michel Wuyts & Louis De Peuter.

Jacob Harrewijn after Jacob van Croes, *The Rubens House at Antwerp ('Maison Hilwerve à Anvers, dit l'hostel Rubens')*, 1684, Collection of the City of Antwerp, Museum Plantin-Moretus – UNESCO World Heritage Site.

Jacques Jordaens, *Cupid and Psyche*, c. 1640–1650, Museo Nacional del Prado, Madrid. © Photographic Archive Museo Nacional del Prado.

Jan Brueghel I, *Flowers in a Wooden Tub*, c. 1607–1608, Kunsthistorisches Museum, Vienna. © KHM-Museumsverband.

Jan Brueghel I and Peter Paul Rubens, *Allegory of Smell*, c. 1617, Museo nacional del Prado, Madrid. © Photographic Archive Museo Nacional del Prado.

Jan Brueghel I, *The Archdukes Albert and Isabella in the Garden of Coudenberg Palace in Brussels*, 1620s, Collection of the City of Antwerp, Rubenshuis. Photo: Michel Wuyts & Louis De Peuter.

Jan Brueghel II, *Allegory of Painting*, c. 1625–1630, JK Art Foundation.

Jan Brueghel II, *Satire of Tulipomania*, c. 1637, Frans Hals Museum, Haarlem. Acquired with support from the Rembrandt Association. Photo: René Gerritsen.

Jan van der Groen, *Van't Hof-gereetschap* from *Den Nederlandsen hovenier*, 1696, Collection of the City of Antwerp, Museum Plantin-Moretus – UNESCO World Heritage Site.

Jan van Kessel I, *Insects and a Sprig of Rosemary*, 1653, Courtesy National Gallery of Art, Washington.

Jan van Kessel I, *Insects with Creeping Thistle and Borage*, 1654, The National Gallery, London, NG6667, Gift from the collection of Willem Baron van Dedem, 2017.

Johannes Bosschaert, *Still life with tulips*, Nationalmuseum, Stockholm. Photo: Nationalmuseum.

Joris Snaet, *Drawing of the Calvinist Temple in the Later Rubens Garden*, 2023.

Joris Snaet, *The Rubens Garden*, 2024 (gatefold).

Matthias Lobelius, *Tulips* in the *Kruydtboeck*, 1581, Collection of the City of Antwerp, Museum Plantin-Moretus – UNESCO World Heritage Site.

Osias Beert, *Flowers in a Glass Vase in a Niche*, 1610–1620, Snijders&Rockoxhuis, Antwerp.

Osias Beert, *Fruit, Confectionery, and Wine in a Niche*, c. 1615–1620, private collection.

Payment to Jaspar the gardener for tending to the orange trees in the *Staetmasse* or settlement of Rubens's estate in 1645, N 1894, Felixarchief, Antwerp.

Peter Paul Rubens, *Peter Paul Rubens, Helena Fourment, and Their Son Fransje*, c. 1635, The Metropolitan Museum of Art, New York (p. 2).

Peter Paul Rubens (atelier), *Peter Paul Rubens, Helena Fourment and Their Son Nicolaas Walking in Their Garden ('The Walk in the Garden')*, c. 1630-1631, Bayerische Staatsgemäldesammlungen – Alte Pinakothek, Munich.

Peter Paul Rubens, *The Artist and His First Wife Isabella Brant in the Honeysuckle Bower*, c. 1609, Bayerische Staatsgemäldesammlungen – Alte Pinakothek, Munich.

Pieter Holsteyn II, *Citrus medica, Erythronium dens-canis* and *Fritillaria meleagris* in *Flores a Petro Holstein ad vivum depicti*, 1640s, RHS Lindley Collections, London.

Pieter Serwouters, *Suikerpeer (Sugar Pear)*, 1648, Collection of the Drents Museum, Assen.

Rembert Dodoens, Carolus Clusius and Jan Christophe Jegher, *Hoppe-cruydt* from the *Cruydt-boeck Remberti Dodonaei*, 1644, Collection of the City of Antwerp, Museum Plantin-Moretus – UNESCO World Heritage Site. Photo: Peter Maes.

Virgilius Bononiensis and Gillis Coppens van Diest, *Map of Antwerp*, 1565, Collection of the City of Antwerp, Museum Plantin-Moretus – UNESCO World Heritage Site. Photo: Michel Wuyts.

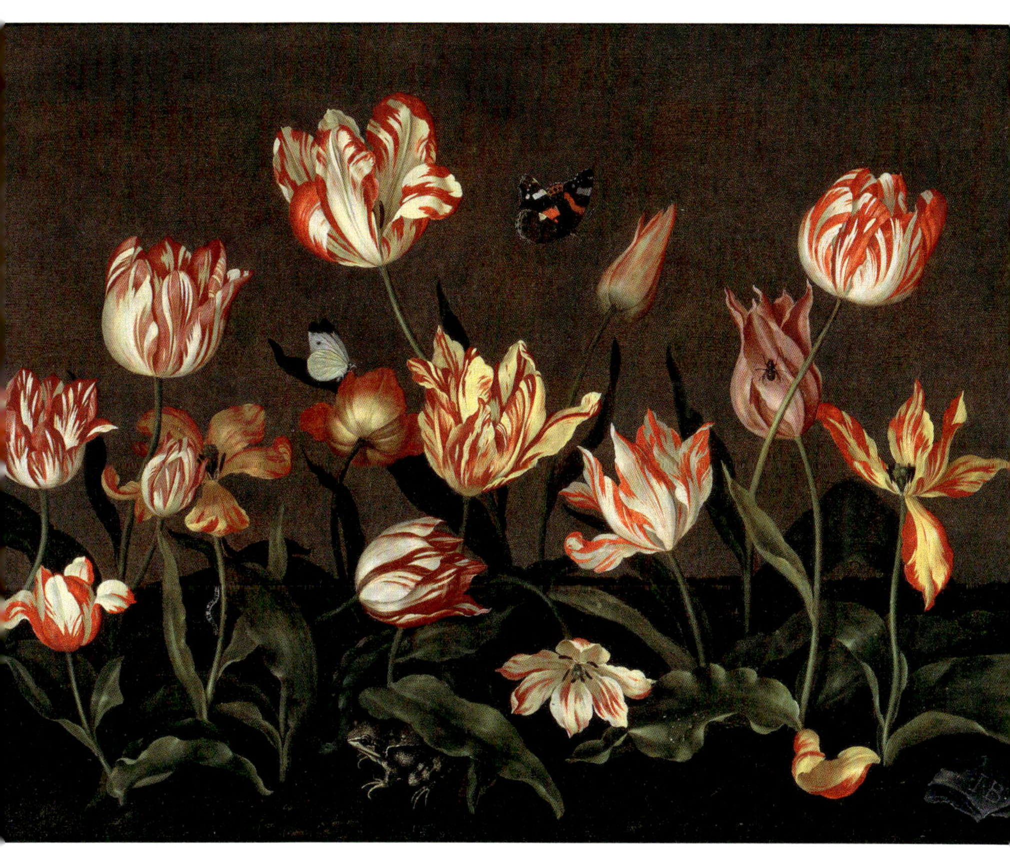

Johannes Bosschaert, *Still Life with Tulips*, Nationalmuseum, Stockholm

This book was published on the occasion of the redevelopment of the garden of the Rubenshuis in 2024, designed by Ars Horti.

Text
Klara Alen

Translation
Sandy Logan

Copy-editing
Derek Scoins

Project management
Sara Colson

Design
doublebill.design

Retouching
Tine Deriemaeker (die Keure)

Printing and binding
die Keure, Bruges

Publisher
Gautier Platteau

ISBN 978 94 6494 159 3
D/2024/11922/66
NUR 654

www.hannibalbooks.be
www.rubenshuis.be

Cover
Jan Brueghel I and Peter Paul Rubens, *Allegory of Smell* (detail), c. 1617, Museo nacional del Prado, Madrid

The American Women's Club of Antwerp is a social, cultural and philanthropic club of international English-speaking women from all over the world. The club celebrated its 95th anniversary in 2024, a rich history of bringing English speaking women in the Antwerp area a place and purpose in a country that, for many, is not their own.

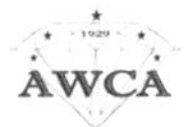

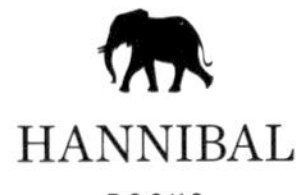

RUBENSHUIS

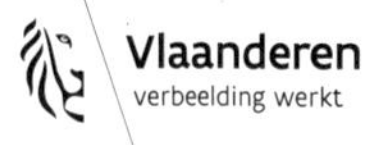